Routledge Introductions to Environment Series

D0139624

Environment and Tourism

Second Edition

Andrew Holden

Routledge
Taylor & Francis Group

LONDON AND NEW YORK

First published 2008
by Routledge
2 Park Square, Milton Park, Abingdon, Oxon, OX14 4RN

Simultaneously published in the USA and Canada
by Routledge
270 Madison Avenue, New York, NY 10016

Routledge is an imprint of the Taylor & Francis Group, an informa business

© 2008 Andrew Holden

Typeset in Times New Roman by Keyword Group Ltd
Printed and bound in Great Britain by TJ International Ltd, Padstow, Cornwall

British Library Cataloguing in Publication Data
A catalogue record for this book is availabe from the British Library

Library of Congress Cataloging in Publication Data
Holden, Andrew, 1960–
    Environment and tourism / by Andrew Holden.—2nd ed.
        p.cm.
    Includes bibliographical references and index.
1. Tourism—Environmental aspects. 2. Tourism—Economic aspects.
3. Tourism—Social aspects.
I. Title.

G155.A1H64 2008
338.4′791—dc22   2007023212

ISBN 10: 0–415–39954–8 (hbk)
ISBN 10: 0–415–39955–6 (pbk)
ISBN 10: 0–203–93762–7 (ebk)

ISBN 13: 978–0–415–39954–8 (hbk)
ISBN 13: 978–0–415–39955–5 (pbk)
ISBN 13: 978–0–203–93762–4 (ebk)

Environment and Tourism, Second Edition

Taking a holiday has become an expected part of lifestyle for many people living in economically advanced societies. However, with the number of international tourism arrivals now surpassing 800 million people per annum and expected to reach over 1.6 billion by 2020, tourism has consequences for the natural environments and communities it comes into contact with. This book offers an integrated account of the problems and opportunities this migration of people presents for nature and communities.

The second edition of *Environment and Tourism* reflects changes in the relationship between tourism, society and the natural environment in the first decade of the new century. Alongside the updating of all statistics, environmental policy initiatives, examples and case studies, new material has been added. This includes two new chapters: one on climate change and natural disasters; and the other on the relationship between tourism and poverty. These themes have direct relevance not only to tourism but are reflective of the wider relationship between nature and society, a thesis that contextualises the book. Tourism is also analysed as an interconnected system, linking the environments of where tourists come from, with the ones they go to. Further issues addressed in the book include tourism's interaction with nature; economic opportunities for conservation; market failure that causes tourism to create environmental problems; environmental management and planning for tourism; environmental ethics; sustainable tourism and ecotourism; poverty and tourism; climate change, natural disasters and tourism.

The book is original in taking a holistic view of the tourism system and how it interacts with the natural environment, illustrating the positive and negative effects of this relationship, and importantly how tourism can be planned and managed to encourage natural resource conservation and aid human development. It will be useful to those studying Human Geography, Tourism and Environment Studies.

Andrew Holden is Director for the Centre for Research into Environment and Sustainable Tourism (CREST) at the University of Bedfordshire. He is a Fellow of the Royal Geographical Society in London. His research and consultancy work has been undertaken in Indonesia, Nepal, Turkey, Russia, Scotland and Cyprus.

Routledge Introductions in Environment Series
Published and Forthcoming Titles

Titles under Series Editors:
Rita Gardner and A.M. Mannion

Environmental Science texts

Atmospheric Processes and Systems
Natural Environmental Change
Environmental Biology
Using Statistics to Understand the
  Environment
Environmental Physics
Environmental Chemistry
Biodiversity and Conservation,
  2nd Edition
Ecosystems, 2nd Edition

Forthcoming:
Coastal Systems, 2nd Edition

Titles under Series Editor:
David Pepper

Environment and Society texts

Environment and Philosophy
Energy, Society and Environment,
  2nd edition
Gender and Environment
Environment and Business
Environment and Law
Environment and Society
Environmental Policy
Representing the Environment
Sustainable Development
Environment and Social Theory,
  2nd edition
Environmental Values
Environment and Politics, 3rd Edition
Environment and Tourism, 2nd Edition

Forthcoming:
Environment and the City
Environment, Media and Communication
Environment and Food

Contents

Series editor's preface to the First Edition   *vii*

List of figures   *xi*

List of boxes   *xiii*

Preface   *xvii*

Acknowledgements   *xxi*

Chapter 1   Introducing tourism   1

Chapter 2   Perceptions of environments for tourism and
             ethical issues   25

Chapter 3   Tourism's relationship with the environment   65

Chapter 4   Tourism, the environment and economics   103

Chapter 5   Environment, poverty and tourism   128

Chapter 6   Sustainability and tourism   148

Chapter 7   The environmental planning and management
             of tourism   171

Chapter 8   Climate change, natural disasters and tourism   210

Chapter 9   The future of tourism's relationship with
             the environment   226

Bibliography   252

Index   268

Series editor's preface to the first edition *Environment and Society titles*

The modern environmentalist movement grew hugely in the last third of the twentieth century. It reflected popular and academic concerns about the local and global degradation of the physical environment which was increasingly being documented by scientists (and which is the subject of the companion series to this, Environmental Science). However it soon became clear that reversing such degradation was not merely a technical and managerial matter: merely knowing about environmental problems did not of itself guarantee that governments, businesses or individuals would do anything about them. It is now acknowledged that a critical understanding of socio-economic, political and cultural processes and structures is central in understanding environmental problems and establishing environmentally sustainable development. Hence the maturing of environmentalism has been marked by prolific scholarship in the social sciences and humanities, exploring the complexity of society–environment relationships.

Such scholarship has been reflected in a proliferation of associated courses at undergraduate level. Many are taught within the 'modular' or equivalent organisational frameworks which have been widely adopted in higher education. These frameworks offer the advantages of flexible undergraduate programmes, but they also mean that knowledge may become segmented, and student learning pathways may arrange knowledge segments in a variety of sequences – often reflecting the individual requirements and backgrounds of each student rather than more traditional discipline-bound ways of arranging learning.

The volumes in this Environment and Society series of textbooks mirror this higher educational context, increasingly encountered in the early twenty-first century. They provide short, topic-centred texts on social science and humanities subjects relevant to contemporary society–environment relations. Their content and approach reflect the fact that each will be read by students from various disciplinary backgrounds, taking in not only social sciences and humanities but others such as physical and natural sciences. Such a readership is not always familiar with the disciplinary background to a topic, neither are readers necessarily going on to further develop their interest in the topic. Additionally, they cannot all automatically be thought of as having reached a similar stage in their studies – they may be first- , second- or third-year students.

The authors and editors of this series are mainly established teachers in higher education. Finding that more traditional integrated environmental studies and specialised texts do not always meet their own students' requirements, they have often had to write course materials more appropriate to the needs of the flexible undergraduate programme. Many of the volumes in this series represent in modified form the fruits of such labours, which all students can now share.

Much of the integrity and distinctiveness of the Environment and Society titles derives from their characteristic approach. To achieve the right mix of flexibility, breadth and depth, each volume is designed to create maximum accessibility to readers from a variety of backgrounds and attainment. Each leads into its topic by giving some necessary basic grounding, and leaves it usually by pointing towards areas for further potential development and study. There is introduction to the real-world context of the text's main topic, and to the basic concepts and questions in social sciences/humanities which are most relevant. At the core of the text is some exploration of the main issues. Although limitations are imposed here by the need to retain a book length and format affordable to students, some care is taken to indicate how the themes and issues presented may become more complicated, and to refer to the cognate issues and concepts that would need to be explored to gain deeper understanding. Annotated reading lists, case studies, overview diagrams, summary charts and self-check questions and exercises are among the pedagogic devices which we try to encourage our authors to use, to maximise the 'student friendliness' of these books.

Hence we hope that these concise volumes provide sufficient depth to maintain the interest of students with relevant backgrounds. At the same time, we try to ensure that they sketch out basic concepts and map their territory in a stimulating and approachable way for students to whom the whole area is new. Hopefully, the list of Environment and Society titles will provide modular and other students with an unparalleled range of perspectives on society-environment problems: one which should also be useful to students at both postgraduate and pre-higher education levels.

David Pepper

May 2000

Series International Advisory Board

Australasia: Dr P. Curson and Dr P. Mitchell, Macquarie University

North America: Professor L. Lewis, Clark University; Professor L. Rubinoff, Trent University

Europe: Professor P. Glasbergen, University of Utrecht; Professor van Dam-Mieras, Open University, The Netherlands

Figures

1.1 The tourism system: an environmental perspective   9
1.2 International arrivals, 1950–2020   21
2.1 Poster – Midland Railway: Cook's excursion to Ireland   37
2.2 Poster – Uganda Railway: British East Africa   38
2.3 Image of Kiribati   40
2.4 'Paradise' is placed under English management   50
3.1 Pre-tourism development: the ideal paradise   79
3.2 Post-tourism development: the disaster   79
3.3 Beach construction alongside the Indian Ocean   80
3.4 Privatisation of beach areas may result in local people being denied access to resources they previously enjoyed   87
3.5 Tourism development may result in a uniformity of structure that fails to reflect local building styles, as here in Lanzarote   97
3.6 The relationship between the natural environment, the local economy and tourism   98
4.1 The 'polluter pays' principle   115
5.1 Tourism arrivals: comparative growth rates   129
5.2 How tourism can aid the reduction of poverty   139
7.1 A hierarchy of national goals   172
9.1 Types of tourist, based on their level of interest in the environment   239
9.2 Shades of green in tourism   240

Boxes

1.1 The development of international tourism to Spain   18

1.2 Examples of investment incentives used by governments   20
to encourage tourism development

1.3 World's top spenders on international tourism in 2005   22
and 1990

2.1 The popularity of the beach   33

2.2 The main characteristics for 'quality' tourism   41

2.3 Percentage of population taking a summer vacation   45
in France by settlement type, 1969–88

2.4 Psychological determinants of demand   47

2.5 A phenomenology of tourist experiences   48

2.6 Cohen's (1972) typology of tourists   51

2.7 The characteristics of psychocentrics and allocentrics   53

2.8 Modes of experience of the destination environment   55

2.9 Tourists rush for kill-a-seal pup holidays   62

3.1 The relationship between society, environment and tourism   71

3.2 The negative environmental consequences of tourism   74

3.3 Coral reefs and game parks: issues of tourism   77
development in Kenya

3.4 The effects of downhill skiing upon the environment   82

3.5 Confrontation over water resources in Mexico   85

3.6 Clashes over resources in Goa, India   86

3.7 Emission factors for comparative modes of transport   94

3.8 Gorillas in Rwanda   99

4.1 Services provided by the natural environment for society   104

4.2 Criticisms of the conventional economics approach to the   105
environment

4.3 The world's top tourism earners in 2005   111

4.4 Proportion of a tour operator's prices that are typically   113
received by the destination

4.5	Hardin's 'Tragedy of the Commons' (1968)	119
4.6	Calculating the 'willingness to pay' of visitors to national parks	123
4.7	Debt-for-nature swaps: Ghana	125
5.1	Millennium Development Goals	131
5.2	Developing countries' increasing share of international tourism receipts	135
5.3	Madikwe Game Reserve, South Africa	137
5.4	Pro-poor tourism in Humla, Nepal	141
5.5	Nepal: tourism for rural poverty alleviation (TRPA) project	143
6.1	The origins of sustainable development	151
6.2	Different approaches to development between the 'dominant world-view' and 'deep ecology'	155
6.3	Guiding principles of sustainable tourism	163
6.4	Integrated rural tourism: Senegal	166
6.5	Indicators of sustainable tourism development	167
7.1	Logging in the Democratic Republic of Congo (DRC)	173
7.2	Uncontrolled inward investment for tourism development in Malaysia	175
7.3	Summary of the costs and benefits of national parks	180
7.4	The use of legislation for protected area status in the Balearic Isles	181
7.5	Zoning in the Great Barrier Reef Marine Park, Australia	186
7.6	Factors that will influence the carrying capacities of tourism destinations	189
7.7	Criticisms of environmental impact analysis (EIA)	194
7.8	Association of Independent Tour Operators (AITO) and Responsible Tourism Initiative	195
7.9	Tour Operators' Initiative (TOI) for sustainable tourism development	196
7.10	Environmental auditing in Finland	198
7.11	Touristik Union International (TUI)	199
7.12	Grecotels: environmental auditing and organic farming	200
7.13	The objectives of codes of conduct for tourism	204
7.14	Code of ethics for the tourism industry – the Canadian example	205
7.15	A code of conduct for host communities	206
7.16	The Himalayan Tourist Code	207
8.1	Climate change: implications for the ski industry of the Alps	216
8.2	Climate threat: Cairngorms, Scotland	217

8.3 The Indian Ocean tsunami and tourism 222
9.1 Characteristics of alternative tourism 233
9.2 Dimensions of ecotourism 234
9.3 Guiding principles of ecotourism 236
9.4 Ecotourism in Belize 245

Preface

This book forms part of a series produced by Routledge on the theme of 'environment and society'. The particular emphasis of this publication is upon the interaction that exists between the environment and tourism. Its main purpose is to provide an introductory text for undergraduate students to the concepts and themes that govern this interaction.

Two simple words, 'tourism' and 'environment', but in the course of this book it is only possible to introduce the reader to the complex issues that lie behind these two terms. Both 'tourism' and the 'environment' represent complicated concepts, and both can be interpreted as intricate systems, where actions taken in one part of the system have consequences for its other component parts. This book takes a holistic approach, in trying to understand the complexity of both tourism and the environment, and the relationship that exists between them. Consequently, the book adopts a multidisciplinary stance to the investigation of this relationship and should be of interest to a range of students studying under the aegis of social science. The perspectives used in this book come from the disciplines of geography, sociology, social psychology and economics, as well as the fields of environment, development and tourism studies.

The term 'environment' is used in its widest sense to incorporate all aspects of human behaviour. Cultural, political, economic and social aspects of the environment are an important part of this book, besides considerations purely of the physical environment. All of these factors affect the way we live, and how as humans we interact with each other, as well as the non-human world. In the last half of the twentieth century, the world has witnessed a faster pace of economic development than ever before. This pace of development has placed a tremendous strain upon the natural resources of the earth, and the functioning of the environmental systems that we as humans rely upon for our survival.

Global warming, ozone depletion, desertification and acid rain are all examples of the types of negative environmental changes that have resulted from, or been accelerated by, human actions. Human rights are also sometimes threatened by development, as humans continue to deny rights to other humans, and people are displaced from their traditional lands and denied access to resources. Subsequently, our interaction with fellow humans and the non-human world presents us with many ethical questions. At the beginning of the twenty-first century, we are increasingly being forced to face ethical issues, if not for the survival of other humans and species, then for our own survival. The realisation that we are part of and not separate from this system that we call 'environment' is beginning to dawn on an increasing number of people.

Changes in society, which can be traced to the Industrial Revolution over two hundred years ago, are also influencing the way we live. Increasing rates of urbanisation in both the developed and developing worlds, and the adoption of an ideology of consumerism as a global creed, are placing increasing demands upon our environment to satisfy our needs and desires. The process of urbanisation has had the effect of removing people from nature, and has presented people with the need to define new notions of community. These changes have developed needs and other wants which, combined with an increased level of prosperity, are increasingly satisfied through consumerism. One form of consumerism, which seems to be an increasingly popular way to meet these needs, is tourism. There are now over 800 million people travelling internationally on an annual basis, which is expected to rise to 1.6 billion by 2020. For many in economically advanced societies, tourism has become an expectation of lifestyle, an experience that is consumed with increasing voracity.

A trend of this expanding demand for tourism is for tourists to go further and further afield from where they live. Vacations have already been sold to tourists for the first flights to space. Many of the world's coastlines and mountain areas have been developed for tourism and Antarctica is now a part of the tourist itinerary. In Spain, the first country to experience mass international tourism in the 1950s, tourism has brought vast economic, cultural and physical environmental changes. However, as will be seen in this book, tourism can be an agent of both positive and negative changes. Recognition of tourism as an agent of change – its growing importance in the world economy and the effects that it can have for the environment – was given at the United Nations General Assembly Special Session (Earth Summit II) in New York in 1997. The basis of this recognition was that tourism must be developed in a sustainable fashion, to ensure the conservation of resources for future generations to make their livelihoods from tourism, just as their parents do now. A fine aim that probably few of us would disagree with, but in reality a concept that represents a diversity of opinion over how it should be achieved, which in turn reveals much about the complexity of the environment in which tourism operates.

Since the first edition of this book, tourism's relationship with the natural environment has in some aspects matured and in other aspects made little progress. Progression can be recognised in the sense that demonstrating a 'green' or environmental profile seems to have become part of the mission of a significant section of the tourism industy, and the concept of sustainability for all its misgivings, has become an integral component of tourism policy and strategy. Tourism has also become part of the debate about global warming, notably in the context of the contribution of aircraft emissions to greenhouse gases (GHGs). New initiatives in tourism development have emerged, notably pro-poor tourism (PPT) and Sustainable Tourism for the Elimination of Poverty (STEP) programmes. The link between poverty and environmental degradation will be an important one to mitigate this century, both for the benefits of human livelihoods, and for the conservation of the natural environment. Unfortunately in other areas, little progress has been made. Disagreement still exists about the meaning of key terminology, e.g. ecotourism and sustainable tourism, making its implementation patchy. Aircraft emissions are still not part of the Kyoto Treaty and the industry continues to fight off the imposition of tax on kerosene for use in aviation. It is this on-going complexity of the interaction between 'tourism' and the 'environment' that this book sets out to explore.

Acknowledgements

The list of people I would like to credit would run into pages and therefore I have purposely avoided a list of names. However, I would like to give special thanks to all of those who have made working in the field of tourism and environmental studies both intellectually stimulating and fun over the years. Thanks also to Tourism Concern, for the permission to reprint the Himalayan code shown in Box 7.15, to Elsevier for the permission to reprint Figure 9.2, and Her Majesty's Stationery Office for the permission to reprint Figures 2.1 and 2.2. Thanks also to Jennifer Page of Routledge for her advice on the practicalities of this second edition. My thanks also to the reviewers of the proposal for the second edition, whose comments were very helpul in guiding the manuscript. A big thanks to my wife, Dr Kiranjit Kalsi, for her support while writing this book.

1 Introducing tourism

- Understanding tourism
- Tourism as a system
- The history of tourism
- Conditions for the growth of tourism

Introduction

To understand the interaction that exists between tourism and the environment, it is necessary to understand the complexity of tourism. Although tourism may seem a relatively simple concept, it is in fact a product of a variety of interacting factors in our home environment, which have consequences for destination environments. Since the 1950s, the impacts of tourism have been experienced more widely, as the numbers of tourists travelling internationally have increased and destinations become more diffuse. This trend has been accompanied by increased amounts of domestic tourism, notably in the rapidly developing economies of Asia and Latin America. This growing demand for tourism is a reflection of changing economic and social conditions in home environments, as much as it is about the physical and cultural characteristics of the environments that tourists travel to. This chapter examines the meaning and complexity of tourism, its history, and how social changes since the Industrial Revolution have shaped contemporary mass participation in tourism.

What is meant by tourism?

The acceptance of 'going away' on holiday, commonly referred to as tourism, as a part of our lifestyle in contemporary western society may lead us to believe that it has always been a feature of people's lives. Yet the word 'tourist' is a fairly new addition to the English language, the word 'tour-ist' (deliberately hyphenated), first appearing in the early nineteenth century (Boorstin, 1992). Boorstin draws a distinction between the arduous conditions undertaken by 'travellers' (a term originating from the French word *travail* meaning work, trouble, torment), such as pilgrims, and the 'tourist', for whom travel has become an organised and packaged affair.

The idea of travel for pleasure, for example, to visit beautiful landscapes as opposed to travel for necessity or to demonstrate religious piousness, is therefore a relatively recent phenomenon. Until the nineteenth century and the advent of the railways, travel was not an easy option, nor were landscapes that we now regard as aesthetically pleasing necessarily regarded in the same way. Yet, today the word 'tourism' has become part of our common language, with over 800 million people per annum travelling internationally (UNWTO, 2006a).

Despite this impressive figure, trying to define what tourism is has proved to be more problematic than might be expected. This difficulty is a reflection of both the complexity of tourism, and the fact that different stakeholders or groups with an interest in tourism are likely to have different aspirations of what they hope to achieve from it, and subsequently hold different perspectives on what it means to them. The stakeholders in tourism include governments, the tourism industry, donor agencies, local communities, non-government organisations (NGOs) and tourists.

Definitions and types of tourism

For the majority of people who possess the financial means to participate in travel for recreational purposes, tourism is an activity that probably little conscious thought is given to beyond recollecting the enjoyment of the last holiday, and deciding where to go to for the next one. Yet the facilitation of this seemingly simple process typically involves the participation of various stakeholders, including national governments, the tourism industry and local communities, all of whom will have their own interests in and expectations of tourism.

Attempts to define tourism are made difficult because it is a highly complicated amalgam of various parts. These parts are diverse, including the following: human feelings, emotions and desires; natural and cultural attractions; suppliers of transport, accommodation and other services; and government policy and regulatory frameworks. Subsequently it is difficult to arrive at a consensual definition of what tourism actually is, as commented upon by many authors of tourism texts (for example, Mathieson and Wall, 1982; Murphy, 1985; Middleton 1988; Bull, 1991; Laws, 1991; Ryan, 1991; Mill and Morrison, 1992; Davidson, 1993; Gunn, 1994; Burns and Holden, 1995; Cooper *et al.*, 1998; Holloway, 1998).

Yet trying to understand the meaning of 'tourism' is important if we are to plan the use of natural resources and manage impacts associated with its development. What all commentators would probably agree with is that tourism involves travel, although how far one has to travel and how long one has to be away from one's home location to be categorised as a tourist is debatable. A convenient definition that overcomes this difficulty is the one proposed by the World Tourism Organization (1991) which was subsequently endorsed by the UN Statistical Commission in 1993: 'Tourism comprises the activities of persons travelling to and staying in places outside their usual environment for not more than one consecutive year for leisure, business or other purposes.'

The preceding definition challenges the commonly held perception that tourism is purely concerned with recreation and having fun. Whilst recreational tourism is the most usual form of tourism, other types of tourism also exist. For instance, Davidson (1993), besides recognising leisure or recreation (in which he includes travel for holidays, sports, cultural events, and visiting friends and relatives) as the main type of tourism, points out that people also travel for business, study (or education), religious and health purposes. Indeed, the origins of tourism lie in travel for reasons of faith, education and health, as detailed later in this chapter.

Although business tourism may initially seem to have little relevance to a text dealing with the interaction between tourism and the environment, it is a particularly important market sector for the economies of many urban environments. Tourism has been purposely used in government policy as a catalyst to aid the regeneration of economically depressed areas of post-industrial cities, such as Baltimore in the USA and Liverpool in the United Kingdom, a theme discussed in Chapter 3. Business travel can be viewed as including travel for the purposes of commerce, exhibitions and trade fairs, and conferences.

From the previously cited World Tourism Organization (1991) definition, it can be inferred that tourism involves some element of interaction with a different type of environment to the one found at home. The consequences of this interaction are commonly referred to as the 'impacts of tourism', and can be categorised into three main types: economic, social and environmental. All these types of impacts can be either positive or negative and are discussed more fully in the course of this book. Recognition of the impacts that tourism can have on a destination environment are noted in the following definition of tourism given by Mathieson and Wall (1982: 1): 'The study of tourism is the study of people away from their usual habitat, of the establishments which respond to the requirements of travellers, and of the impacts that they have on the economic, physical and social well-being of their hosts.'

Besides referring to the impacts of tourism, Mathieson and Wall's definition adds a further dimension to the concept of tourism by introducing a behavioural dimension, that is, the 'study of people away from their usual habitat'. Given that tourism would not exist without tourists, the motivations of tourists and the effect of their behaviour on the environments of destinations are of interest to social psychologists, sociologists and anthropologists. The last word of this definition, 'host', implies an invitation from people who are happy to receive tourists. This term has received increasing criticism from academics, NGOs and the more socially aware quarters of the tourism industry, as levels of cultural and environmental awareness have grown since the early 1980s. It is now recognised that in some cases tourism is something that is tolerated or even forced upon communities as opposed to necessarily being welcomed.

Introducing the dimension of resource usage into tourism, Bull (1991:1) suggests that: 'It [tourism] is a human activity which encompasses human behaviour, use of resources, and interaction with other people, economies and environments.' This definition draws attention to the fact that natural and cultural resources are the focus of much of tourism, and these are 'used' in various ways, including for pleasure, financial gain and economic development.

The idea of wealth creation through the use of the natural and cultural environments for tourism has received increased international attention since the success of General Franco's policies on tourism development in Spain in the 1950s, as discussed later in this chapter. Today, the focus of using tourism for development has special significance for many developing countries, including countries in Africa, Asia and Latin America. However, the extent to which environmental resources should be used for tourism and traded off for economic development may be debatable and contentious, sometimes raising ethical and political questions. Similarly, ethical questions can be raised over tourism's 'interaction with other people', and the extent to which it is beneficial for local or indigenous communities. These are themes which are developed in the course of this book.

Another way to think about tourism is from the perspective of the tourist, emphasising the experiential dimension. For example, Franklin (2003: 33) defines tourism as: 'an attitude to the world or a way of seeing the world, not necessarily what we find only at the end of a long and arduous journey'. In this definition, emphasis is subsequently placed upon individuals constructing their own meaning of tourism instead of it existing as a defined entity. Similarly, as is discussed in Chapter 2, people may mentally construct their own environments, rather than it existing purely as a defined and real entity.

This brief analysis of various definitions of tourism demonstrates its complexity, and highlights that it is about much more than simply 'going on vacation or holiday'. Tourism is based upon the economic and social processes and changes that are occurring in the environments of the societies where tourists come from. Its development in destinations, focused upon the use of natural and cultural resources, will have economic, environmental and cultural impacts. It may also be interpreted as something that we seek to measure or something that we experience.

What is the 'tourism industry'?

According to the World Travel and Tourism Council (WTTC) (2007), travel and tourism contributed directly and indirectly to the global economy in 2006:

- 10.3 per cent of Gross Domestic Product;
- 234.3 million jobs; and
- 8.7 per cent of total employment.

Although these are impressive statistics and reference is often made to tourism being one of the world's largest industries, trying to define what is meant by the 'tourism industry' is extremely difficult: 'The problem in describing tourism as an "industry" is that it does not have the usual production function, nor does it have an output which can physically be measured, unlike agriculture (tonnes of wheat) or beverages (litres of whisky)' (Lickorish and Jenkins, 1997:1).

They also add that the vague and dispersed nature of the tourism industry has made it difficult to evaluate its impact upon the economy relative to other economic sectors. Similarly, in destinations where tourism development has taken place and environmental problems have arisen, it is not always easy to disaggregate tourism's contribution to these problems from the contributions of other economic sectors.

Murphy (1985) suggests that a tourism industry does not exist because it does not produce a distinct product. He points out that certain industries such as transport, accommodation and entertainment are not exclusively tourism industries, for they sell these services to local residents as well. Another key difference between tourism and other industries is that it is the consumer who travels to the 'product' and not vice versa. The major inference of this last point is that the natural and cultural resources of destinations can be treated as a form of product, the qualities of which may be traded in the marketplace. It is these characteristics of environments that create expectations in tourists and form a vital part of their experience.

Yet within tourism, practices are adopted that are familiar to those of industrial production. For instance, a common expression used in connection with contemporary tourism is 'mass tourism'. Mass tourism involves tour operators compiling a standardised package, at the very minimum usually comprising transport and accommodation, which is then sold into the marketplace *en masse* to thousands of consumers. This 'package holiday', which relies on mass consumption and sales to keep the prices low, displays the characteristics of 'Fordist' production. This is a term used to describe the mass conveyor-belt techniques pioneered by Henry Ford in the motor car industry in the early twentieth century, which paved the way for mass car ownership, by keeping production costs low. Although the development of mass tourism has meant that millions of people have had the opportunity to travel to different countries, its development in destinations is often associated with environmental problems, such as pollution and a loss of local culture. Commenting on mass tourism, Poon (1993: 4) writes:

> Mimicking mass production in the manufacturing sector, tourism was developed along assembly-line principles: holidays were standardised and inflexible; identical holidays were mass produced; and economy

of scale was the driving force of production. Likewise, holidays were
consumed *en masse* in a similar, robot-like and routine manner, with a
lack of consideration for the norms, culture and environment of host
countries visited.

Another comparison between tourism and other industries is somewhat
polemically given by Krippendorf (1987: 19): 'The timber industry
processes timber. The metal industry processes metal. The tourist
industry processes tourists.' Yet, although the existence of a tourism
industry is debatable, there are definite types of businesses that are
specifically orientated to providing the services that meet the needs of
tourists, including travel agents and tour operators, airlines and the
international accommodation sector.

Although the industry consists primarily of small and medium size
enterprises, some organisations operate on a global scale, e.g. Sheraton
and Hilton International Hotels, whilst some tour operators have grown
into major transnational businesses and are now listed on the stock
exchange, e.g. Touristik Union International and Carlton Travel.
Similarly, some airlines have become transnational businesses, with
companies such as American Airlines and British Airways seeking
strategic alliances to increase their global market share.

In summary, the 'tourism industry' cannot be likened to other industries
as it does not produce a single identifiable product and neither are many
of its services used exclusively by tourists. Essentially when the term
'tourism industry' is used, it refers to an amalgam of different businesses
and organisations, connected by the common factor of providing services
in some capacity to tourists.

Tourism as a system

Another approach to understanding tourism is to think of it as a system,
incorporating not only businesses and tourists, but also societies and
environments. Some authors interpret these different components of
tourism as being interlinked, thereby forming a 'tourism system'. For
instance, Gunn (1994) advocates that tourism should be interpreted as a
system, adding that every part of tourism is related to every other part,
and that no manager or owner involved in the tourism system has
complete control over his or her own destiny. For this reason, managers
involved in any part of this system need to understand its complexity and

possess a holistic cum reductionist view of their business operations. The decisions and actions taken by businesses will have consequences for other components of the system. For instance, the decision of a tour operator to axe a particular destination from their schedule will have economic and social consequences for the businesses and local community in that destination that rely upon their trade. Similarly, the predicted climatic changes associated with global warming pose a significant threat to existing patterns of tourism demand, a theme that is explored in Chapter 8.

According to Stephen Page (1995), the advantage of a systems approach is that it allows the complexity of the real life situation to be accounted for in a simple model, demonstrating the linkages of all the different elements. Mill and Morrison (1992) use the analogy of a spider's web to illustrate the relatedness of different parts of the tourism system, in which touching one part of it induces a ripple effect throughout the web. According to Laws (1991), the advantages of interpreting tourism as a system are that it avoids one-dimensional thinking and facilitates a multidisciplinary perspective. Such an approach is beneficial *vis-à-vis* a topic that can be interpreted from a range of disciplinary perspectives including economics, psychology, sociology, anthropology and geography. The components of the tourism system modified from Laws (1991) to include a heightened environmental perspective are shown in Figure 1.1.

This model incorporates a range of different elements which together form the tourism system. Important inputs to the system from an environmental perspective include natural and human resources, the use of which are encouraged by both consumer demand in the market system for tourism, and government policy aimed at increasing entrepreneurial activity and inward investment in the sector. Within the overall system, three distinct subsystems are recognisable – tourism retailing, destination and transport – all of which overlap and are interrelated. Incorporated in these subsystems are the businesses that have been developed to cater primarily for tourists, such as tour operators, international hotel companies, global airlines and locally owned tourism enterprises. Within the destination subsystem, the importance of natural and cultural attractions is emphasised as the basis for attracting tourists.

The outputs of the system, which may alternatively be expressed as outcomes, suggest that tourism will bring environmental and cultural changes. These changes illustrate the dichotomy of tourism in the sense

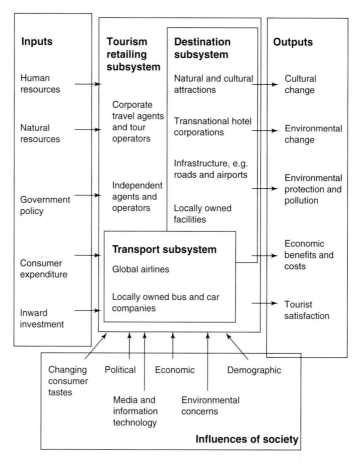

Figure 1.1 *The tourism system: an environmental perspective*
Source: After Laws (1991)

that they can be either positive or negative. Tourism can both conserve and pollute the physical environment. It can also bring positive and negative cultural changes, such as employment opportunities for women or result in women being forced into prostitution. Similarly it can create economic opportunities for communities but can also result in an economic overdependence on tourism and encourage price inflation. Another output of the system, which is essential for ensuring the profits of enterprises based upon tourism and helping to secure the economic benefits desired by governments, is tourist satisfaction.

Finally, the tourism system is subjected to a range of influences exerted by changes in society. These may be classified, using a term borrowed from Poon (1993), as 'framing conditions'. Within the context of this model, the term applies to those conditions in society that influence the working of the system. For instance, in the 1990s, Poon observed changing consumer tastes, being typified by the emergence of the 'new tourist'. These tourists display characteristics of being more environmentally aware, independent, flexible and quality conscious than the tourists who constitute the mass market. Subsequently, the tourism retailing subsystem must adjust its product to facilitate this new market segment, and local governments and municipalities must plan and develop their destination in a way to attract this market segment.

Economic, technological and political changes also influence participation in tourism, either positively or negatively. For example, positive developments that facilitate travel include: increasing economic prosperity; political freedom to travel; cheaper, faster and more comfortable transport options; and the spread of the internet. Conversely, negative changes such as economic recession and poverty, or terrorist acts like September 11th 2001, threaten tourism demand. By threatening demand, the livelihoods of many people in tourism destinations are also made vulnerable. These examples help to illustrate the complexities and interlinkages of the tourism system.

THINK POINT

Why would it be more appropriate to think of tourism as a 'system' rather than as an 'industry'?

The growth in demand for tourism

Although tourism is a feature of global society, it is only relatively recently that it has become an activity of mass participation. The growth in demand for tourism is a reflection of a range of changes that have occurred in society, particularly since the onset of the Industrial Revolution. This section of the chapter outlines the history of tourism and reasons for its growth, continuing to describe a type of tourism which was particularly significant in the latter half of the twentieth century, namely international mass tourism. The consequence of mass participation in international tourism, from an environmental viewpoint, is that an increasing number and variety of natural environments are being exposed to tourism, with a range of consequences.

Pre-industrial tourism

Tourism is not something that happens by chance but is an activity that has developed as a consequence of the types of societies in which we live. Certainly the tendency for people to live in urban areas would seem to increase the propensity for tourism. Although we tend to think of mega-cities as very much a phenomenon of contemporary times, large cities have existed in ancient history, for example, Carthage at its fall in 146 BC had a population of 700,000 and Augustan Rome a population of 1 million (Goudie and Viles, 1997). Similar to the physiological needs that motivate many people to leave cities in summer in contemporary times, the desire of the Romans to escape the heat of Rome led them to travel to seaside and mountain villas (Holloway, 1998). There was evidence of the development of a hierarchy of resorts, possessing distinct types of cultural environments, and attracting different market segments. Holloway (ibid.: 17) comments:

> Naples itself attracted the retired and intellectuals, Cumae became the resort of high fashion, Puteoli attracted the more staid tourist, while Baiae, which was both a spa town and a seaside resort, attracted the down-market tourist, becoming noted for its rowdiness, drunkenness and all-night singing.

The extent of this tourism development in Roman times was to remain unmatched until the development of the French Riveria, nearly 2000 years later (Davidson and Spearitt, 2000). When travelling to resort areas, wealthy Romans would rest at their own private villas *en route*, which although only used for three or four nights per annum were fully staffed by servants and elegantly furnished, whilst commoners rested in tavernas built by farmers on the highway between Rome and the coast (Eadington and Smith, 1992). Long-distance travel was also facilitated in the Roman Empire by the development of a transport infrastructure, the need to use only one currency whilst travelling the Empire's length from Syria in the east to Hadrian's Wall in Britain in the west, and the requirement to speak only Latin.

After the collapse of the Roman Empire in the west in the fifth century AD and the onset of the Middle Ages, travel became more difficult, and from the evidence of historical records was very limited. Travel was arduous, mostly undertaken out of a necessity to trade, or to prove one's religious devotion through pilgrimage. However, the early seventeenth century was marked by the emergence of the 'Grand Tour', a direct outcome of the freedom and quest for learning heralded by the

Renaissance, a period marked by a rediscovery of the classical teachings of the civilisations of Rome and Greece. Holloway (1998) links the establishment of the Grand Tour to the reign of Elizabeth I, when young men seeking positions at court were encouraged to travel to the Continent to finish their education. As Brendon (1991:10) comments: 'It [The Grand Tour] has begun, during Queen Elizabeth's reign, as a refined form of education, a school to "finish" patricians by giving them first hand experience of classical lands.' Essentially the Grand Tour consisted of a tour of European culture for aristocratic young men (the absence of references to women in the accounts of the Grand Tour is noticeable) and the young aristocrats of Britain, France, Germany and Russia were particularly involved in it. The influence of the Grand Tour on subsequent attitudes to travel was notable, as Towner (1996: 96) comments: 'The Grand Tour, that circuit of western Europe undertaken by the wealthy in society for culture, education, health and pleasure, is one of the most celebrated episodes in the history of tourism.'

For the first time since Roman times, foreign environments were seen as being pleasurable, stimulating and educative. The Grand Tour included visiting the major cultural centres of Europe, and lasted from the beginning of the seventeenth century through to the onset of the Napoleonic Wars, in the first part of the nineteenth century. The tour lasted an average of three years and the gentleman would be accompanied by a private tutor. According to Gill (1967), one of the most notable tutors was Adam Smith, the eminent economist and the author of the seminal *An Inquiry into the Nature and Causes of the Wealth of Nations* (1776), in which he advocates trade liberalisation and free market economics. Although the Grand Tour is predominantly viewed as an aristocratic activity, Towner (1996) suggests that this impression is probably a consequence of the likelihood that the written records of the most prominent people are the most probable ones to have survived. In Towner's view, social participation was likely to have been wider than just the aristocracy, with a possible figure of 15,000–20,000 British tourists partaking in the Grand Tour when it was at its zenith in the mid-eighteenth century.

The development of a travel culture during this period was supported and aided by the appearance of guide and travel books such as William Thomas's *The History of Italy* (1549). Health aspects also added another dimension to the Grand Tour; for instance, Montpellier in France developed as a place to visit to counteract the effects of consumption, and destinations in the French Riviera also emerged as sites for health cures.

Nash (1979) describes how Nice became established as a destination for health tourism in the eighteenth century. By the nineteenth century there were accounts of the town filling with visitors from England escaping the winter, although as Nash points out, by this stage healthy tourists probably outnumbered the infirm. It is interesting to reflect how the purpose of tourism affects the seasonal pattern of arrivals. Today, in contrast to the earlier winter visitors Nice experiences its high season in summer, as tourists arrive primarily for recreational purposes in contrast to the earlier health-driven winter visitors.

Travel for the purposes of amusement and pleasure, to enjoy the cultures and social life of cities like Paris, Venice and Florence, also progressively became part of the Tour, and by the end of the eighteenth century this custom was well ingrained (Holloway, 1998). From the mid-eighteenth century, a much greater emphasis also began to be placed upon the viewing of nature, emphasising its spiritual and romantic qualities (Towner, 1996). Typically this entailed the viewing of 'wildscape', such as mountain areas, where there was little human alteration of the landscape. This shift in emphasis was a culmination of the effects of the Industrial Revolution, urbanisation and the Romantic movement, as detailed in the next chapter.

A further development in tourism during the time of the Grand Tour was an increased popularity in the spa towns of Europe, for the first time since the decline of the Roman Empire. Based upon an association between the medicinal qualities of taking the spa waters and good health, spas represented an early type of health tourism. During the eighteenth century spa towns such as Bath in England, Vichy in France and Baden-Baden in Germany reached the height of their popularity. Whilst at its most basic a spa consists of a pump room for water and a bath, this was often followed by the development of a range of other activities such as assembly rooms, ballrooms, gambling casinos and high-class brothels (Hobsbawm, 1975), which Walton (2002) suggests formed their real *raison d'être*. The popularity of spa towns also involved a cultural dimension, their cosmopolitan clientele allowing people to bid for status through fashion and achievement, rather than hereditary lineage or known office. However, by the end of the eighteenth century, spa towns were turning primarily into residential and commercial centres, as coastal areas grew in popularity for tourism. The eventual demise of the Grand Tour in the early nineteenth century is usually attributed to the outbreak of the Napoleonic Wars.

Creating the conditions for the growth of contemporary tourism: the influence of industrialisation and urbanisation

The Industrial Revolution, the origins of which can be traced back to the mechanisation of cotton and wool production in the north of England in the last quarter of the eighteenth century, brought great economic and social changes in society. The extent of these changes led the eminent historian Hobsbawm (1962) to regard the Industrial Revolution as probably the most important event in world history. A major change was the urbanisation of societies in America, Australia and Europe, as people shifted from living in predominantly rural environments to urban ones. For example, from having populations of fewer than 50,000 in 1850, the populations of San Francisco in the USA and Melbourne in Australia had risen to 364,000 and 743,000 respectively by 1891 (Soane, 1993). In total, the number of people living in towns of over 100,000 people in North America and Europe increased from 14,800,000 in 1850 to 80,800,000 by 1913 (ibid.).

Not only did this rural–urban migration have the effect of distancing people from direct contact with nature, it also led to changes in established community structures. Subsequently, perceptions of what were desirable landscapes began to change, and for some sociologists the dominance of the city as a place in which to reside became associated with the isolation of the individual. In the next chapter these themes are discussed within the context of changing perceptions of landscapes and the reasons to explain why people travel.

The Industrial Revolution also made society much more time-conscious than it had been previously. In agricultural society the pattern of labour had been determined by the seasons; now it became highly structured around the need to keep industrial production functioning, with man, woman and child often working a 6-day and 70-hour week (Clarke and Critcher, 1985). Drinking whilst at work and breaking off from work to attend to domestic affairs, which were previously not unfamiliar work practices, were aspects of work life that factory owners could not tolerate. The interaction of people when performing tasks and enjoying leisure together, encompassed within the same spatial area which typified work and leisure practices in pre-industrial society, were essential for establishing agricultural communities. During the Industrial Revolution, work and leisure became highly differentiated, along temporal and spatial zones. This pattern is reflected in contemporary tourism, as we take defined periods of time off work, and travel long distances away from our home environment to other destinations.

Vacation or holiday leave from work is a prerequisite for recreational tourism, and the need for time free of work obligations, a time in which to 're-create' oneself, was recognised in government policy by the end of the nineteenth century. For example, in England the passing of the so-called 'Bank Holiday Acts' in 1871 and 1875 offered a four-day statutory holiday (Towner, 1996). By the twentieth century, government legislation for paid vacation periods from work had became common to most industrialised countries, as in the case of France in 1936, when 12 days of paid vacation became mandatory in all enterprises (Dumazedier, 1967).

Another major change in society was a technological advance in transport, based upon the steam-engine, notably the development of the railways and steamships. Until the nineteenth century, travel was largely dependent upon the horse and carriage, which made journeys arduous: for example, in the United Kingdom a journey of approximately 640 kilometres between London and Edinburgh took 10 days by horse and carriage (Holloway, 1998). In the latter part of the nineteenth century, the development of railway systems and passenger steamship services made travel considerably easier. The railways were also a major force in eroding localism and removing barriers to mobility, bringing together different regions of countries, and ultimately different countries. The use of this advanced technology necessitated a much higher level of investment than the stage coaches, subsequently necessitating the sale of large quantities of passages to secure a return upon one's investment. This meant that railway and steamship companies were required to be active in attracting a diverse usage of their services, including for recreation and tourism.

The impact of technological development on the growth of tourism continued in the twentieth century, with the spreading popularity of the motor car and the advancement of the jet engine, making travel easier and quicker. The car offered people greater control over their own travel arrangements, while the provision of good road infrastructures in many countries has meant that going on holiday by car has become a popular option. The development of jet-powered aircraft has led to travelling times between countries being dramatically reduced, facilitating the access of tourists to foreign countries and increasing the propensity for international travel. This propensity has been increased since the 1990s with the deregulation of air travel and the subsequent growth in budget and low-cost airlines.

In the twenty-first century, the role of information technology already has, and will increasingly have, an important influence upon tourism demand. By the early 1990s, not only was it possible to find out the price, availability and location of a holiday spot, but the technology was already available for clients to take a visual tour of the hotel they would be staying in and the rainforest they would be walking in (Poon, 1993). Information technology has a key role to play in bringing together the demand and supply elements of tourism, and the booking of flights, hotels and other auxiliary tourist services on the internet has become commonplace for many people, facilitating a trend towards individual travel and away from mass travel.

Alongside technological advancement, the growth of a mass participation in tourism has also been facilitated by the emergence of tour operators. Three of the best-known names in the tour operating and travel agency sector are Thomas Cook, Lunn Poly and American Express, all of whom developed their tour-operating businesses in the nineteenth century. Thomas Cook organised his first fee-paying trip in 1841 as secretary of the Midland Temperance Association, with 570 members travelling from Leicester to Loughborough (Holloway, 1998). As Page (1999) remarks, it is rather ironic that the package tour now thrives on an image of sun, sea, sex and booze. By the 1860s Cook had already developed tours to Europe and America, and in 1869 offered the first escorted tour to the Holy Land (Boorstin, 1961). In his first nine years of business, Thomas Cook handled more than one million customers. By the end of the nineteenth century, Sir Henry Lunn was organising trips to the European Alps for British people interested in winter sports, whilst in America the American Express Company had begun to offer Americans help in securing railroad tickets and hotel reservations.

The trend towards mass participation in travel has also been encouraged by a reduction in the real price of travel, with economies of scale becoming possible for suppliers, as more people wished to travel. The reduction in the real price of travel is illustrated by the cost of transatlantic travel between the United States of America and Britain. At the beginning of the twentieth century, to cross the Atlantic by passenger ship was a very rare and special event, restricted to a very few, and subsequently expensive. For instance, a first-class ticket on the ill-fated passenger *Titanic* in 1912 cost US$3000 or approximately US$127,000 in 1990s prices, whilst a third-class ticket cost US$40 or US$1,696 in 1990s prices (Ezard, 1998). Approximately 100 years after the ill-fated voyage of the *Titanic*, a period during which incomes have risen manifold, a typical discounted air fare between London and New York costs a few hundred dollars.

Two phases of mass participation in tourism are observable, the first having taken place in the nineteenth century, fuelled by a rapidly increasing national income per head and the development of the railways. Primarily a mass participation in domestic tourism, it involved the movement of thousands of working-class people from the city areas to the coast, leading to the growth of seaside resorts in northern Europe, as is detailed in Chapter 2. This trend for domestic tourism continued for the first part of the twentieth century, to be succeeded by a second phase of mass tourism occurring after the Second World War, this time having an international perspective.

Mass participation in international tourism

The trend towards a mass participation in tourism was initially most evident in Europe, owing to a combination of economic, social and geographical factors. The amalgam of increasing prosperity, mobility and growth of a tourism industry, combined with the close geographical proximity of nation states, meant international tourism was no longer the preserve of a small elite. Subsequently, although there was a post-Second World War increase in participation in tourism in most developed countries of the world, the geographical size of countries like the United States of America and Australia dictated that tourism remained largely domestic, even though there existed an increased propensity for overseas travel, for example, from the United States to the Caribbean. Thus, this account should be understood in the context of its bias towards Europe, and it is important to consider that increased participation in tourism was also taking place in other parts of the world.

This trend towards mass participation in international tourism in Europe was marked by the beginnings of the demise of many of the northern European seaside resorts. The origin of this second wave of mass tourism was initiated by the movement of thousands of tourists from the United Kingdom to Spain, ultimately leading to the development of the western Mediterranean coastline for tourism. A range of factors explain the flow of tourists to Spain, including the following: an increasing level of disposable income from the late 1950s; a surplus of Second World War aircraft which could be used to provide cheap transport from the UK; the emergence of tour operators, e.g. Vladimir Raitz and the Horizon Travel Company, who promoted Spain as a destination; the encouragement of tourism development by the Spanish dictator General Franco; and the availability of cheap land in Spain for hotel development (see Box 1.1). The subsequent availability of cheap package holidays to Spain brought foreign travel within the reach of the working-class or proletariat for the first time, the vast majority having previously had no

Box 1.1

The development of international tourism to Spain

Although Spain is the country that is probably associated most closely with international mass tourism, its development as a destination for millions of tourists is recent. Spain was never an established part of the Grand Tour, its landscape and culture not regarded as being particularly attractive, and its geographical position rendered it fairly inaccessible. It was not until towards the end of the nineteenth century, with the development of the Romantic movement in northern Europe, that the wild landscape of Spain combined with its medieval and Moorish culture began to be considered attractive by an elite group of travellers, and Spain began to develop an image as being exotic. Its geographical remoteness combined with a lack of infrastructure development, the civil war of the 1930s and two world wars in the first half of the twentieth century, meant Spain retained an exotic image even until the 1950s.

In the 1950s, today's heavily developed tourism areas of the Costa Brava and Costa del Sol began to receive foreign visitors. So rare were foreign visitors in post-war Spain that their movements were monitored by the Civil Guard, a legacy of foreigners who fought in the Spanish Civil War on the side of the Republicans against General Franco. The consequences of tourism on parts of the coast have been noticeable. For example, in 1955 Torremolinos (a name now synonymous with large-scale tourism development) was a poverty-stricken fishing village, where villagers grafted a hard living from the land and sea. The first foreign visitors to arrive in Torremolinos were wealthy foreigners who did not want to walk in the Swiss Alps or sojourn on the French Riviera. The style of the first hotels reflected a luxurious peasant style and by the 1960s Torremolinos had become a highly fashionable resort. Beachfront property increased in price twentyfold in two years, and a villa worth £1,000 in 1955 was sold as a site

Box 1.1 continued

for a hotel in 1963 for £146,000. Torremolinos's popularity meant increasing numbers of visitors arrived on shuttle buses from expanding airports, which led to increased hotel development. Ultimately the rich and the artistic types deserted Torremolinos and went to nearby enclaves at Marbella or as far afield as Bali and Morocco.

Apart from illustrating a typical cycle of resort development, with a destination area being discovered by a wealthy elite leading to rapid unplanned tourism growth determined by market forces, Torremolinos also illustrates the economic opportunities to be gained from tourism. General Franco encouraged the growth of tourism, based upon a need to gain political recognition of a regime that was viewed with international suspicion, and also as a means of attracting foreign exchange and modernising other sectors of the Spanish economy. Spain was marketed as an exotic, low-cost holiday destination, distinct from its chief catchment area of northern Europe. The growth of tourism to Spain was also fuelled by the development of the jet engine, bringing Spain closer to countries of northern Europe in terms of travelling time, and the expansion of the tour-operating industry in northern Europe. Subsequently, Spanish tourism has undergone a major expansion, international arrivals increasing from 700,000 visitors in the early 1950s, to 4 million by 1959, 40 million by the early 1980s, to 56 million in 2005.

Sources: Moynahan (1985); Pi-Sunyer (1996); Barke and Towner (1996); World Tourism Organization (1999); United Nations World Tourism Organization (2006).

opportunity for international travel, beyond the experiences of working-class men as soldiers in the Second World War. Although this trend may have been initiated in the United Kingdom, it was a model quickly followed in other northern European countries as their citizens desired to experience the warmer coastal environments of the south of Europe.

The success of Spain in using tourism as a catalyst for economic growth and development has since encouraged governments of other countries to develop tourism as a part of economic policy. Governments can offer a range of financial incentives to encourage tourism development, including the giving of grants and loans at beneficial interest rates for tourism development, relaxation of import duties, and tax breaks. Examples of actual incentives that have been used to encourage tourism development are shown in Box 1.2.

Box 1.2

Examples of investment incentives used by governments to encourage tourism development

Greece: Incentives were made available for the country's economic and regional development through Law 1262 passed in 1983. Tourism is considered a 'productive investment' and incentives are available for the construction, extension and modernisation of hotels (up to 300 beds), winter sports facilities, spas, tourist apartments, and for the renovation of traditional houses into hostels or hotels. Typical financial incentives include grants, preferential grants, interest rate subsidies, loans, fiscal incentives and tax allowances.

Malaysia: Reliance is placed upon fiscal incentives rather than grants or loans. Companies may be given 'tax holidays' of up to five years providing the development meets certain criteria, such as the number of employees generated, and the geographical locality of the investment. Tax credit of up to 100 per cent is also given on capital expenditure incurred during the first five years of the operation of a tourism facility.

Source: Bodlender and Ward (1987)

A combination of changing economic and social conditions, entrepreneurial activity, and the active promotion of tourism development by national governments has facilitated a continued positive growth in tourism demand. The actual growth in international tourism arrivals for the period 1950–2006, and the projected growth to 2020 is shown in Figure 1.2. As is evident from Figure 1.2, the actual growth in international tourism since 1950, and the projected growth to 2020 are strong. The recorded number of international arrivals (to the nearest million) in 1950 was 25 million, in 1980 this had risen to 270 million, by 2000 it had reached 687 million, and by 2020 it is projected to be 1602 million (World Tourism Organization, 1998a; UNWTO, 2006a).

This rapid growth in international tourism, combined with an expected even larger number of

THINK POINT

Describe the changes that have occurred in the environments of societies where tourists originate from, which help to explain the growth in demand for international tourism.

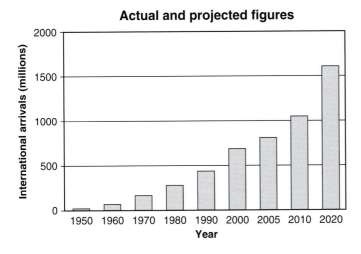

Actual and projected figures

Figure 1.2 *International arrivals, 1950–2020*

arrivals as a consequence of domestic tourism, places pressures upon natural environments and resources. It is not possible to facilitate the visitation of hundreds of millions of peoples to destinations without the development of facilities to accommodate their needs. Consequently, the ability to understand, plan and manage tourism will be decisive in deciding the degree to which its relationship with the environment is held to be positive or negative.

Tourism flows: the influence of climate and nature upon demand for recreational tourism

The importance of economic development as a basis for generating tourism demand is reflected in the ranking of order of countries on the criterion of 'expenditure on tourism', as shown in Box 1.3. Ranked in descending order based upon total expenditure on international tourism for 2005 and 1990, two trends are evident. The first being the dominance of developed countries in the top six places; the second being the growing prominence of those countries with rapidly expanding economies, notably China and to a lesser extent the Russian Federation. In 1990, China was ranked only 40th in terms of tourism expenditure, but by 2006 it had risen to 7th place. Alongside rapid economic growth, the reduction of political restrictions on outbound travel is instrumental in explaining this increased expenditure on tourism by the Chinese. Similarly, the opening of the Russian economy to market forces and the lifting of restrictions on travel after the collapse of communism in 1991 are major factors to explain Russia's rise from 23rd position in 1990 to 10th in 2005.

However, whilst economic development is an important factor for generating tourism demand, the attraction of nature plays a significant

Box 1.3

World's top spenders on international tourism in 2005 and 1990

Country	Actual expenditure on international tourism in 2005 (US$ billion)	Rank in 2005	Rank in 1990
Germany	72.7	1	2
United States	69.2	2	2
United Kingdom	59.6	3	4
Japan	37.5	4	3
France	31.2	5	6
Italy	22.4	6	5
China	21.8	7	40
Canada	18.4	8	10
Russian Federation	17.8	9	N/A
Netherlands	16.2	10	9

N/A=Not available

Source: UNWTO (2006a)

role in deciding where people go to. The world's most numerically significant flow of tourists, approximately 120 million, is from northern Europe to the Mediterranean (WTO, 2003). This is reflected in the economies of many countries and regions of the Mediterranean Basin in which tourism is a significant component. Expressed in terms of tourist flows, there are six major patterns that account for nearly a quarter of all international tourist arrivals. Ranked by approximate numbers of arrivals, these are: (1) northern Europe to the Mediterranean (120 million); (2) North America to Europe (23 million); (3) Europe to North America (15 million); (4) North East Asia to South East Asia (10 million); (4) North East Asia to North America (8 million); and (5) North America to the Caribbean (8 million) (WTO, 2003).

A critical factor to explain the flows of tourism between northern Europe and the Mediterranean and between North America and the Caribbean is the attractiveness of the climate of these destination regions. As Williams (1997) points out, it is the summer beach and the sunshine holiday which has propelled the Mediterranean into pole position in world tourism. The flow between North East Asia to South East Asia also has a large sun, sea and sand component, although business travel and visiting friends and relatives (VFR) are also key components of this market (WTO, 2003.)

However, although these patterns are highly significant, international tourism has expanded to include a variety of destinations and different environments. As Poon (1993) observed in the early 1990s, tourists are demanding the experience of new cultures, physical environments and activities. This has led to indigenous cultures and special ecosystems such as rainforests, coral reefs and polar areas becoming foci of tourism. With technological advancement it would seem that no environment is too remote or inaccessible for tourism. The peripheries of tourism continue to be stretched further afield, now including outer space. Already one American company 'Zeaghram Space Voyages' has sold more than 250 places at £60,000 each to visit space, and plans are already developed for the construction of a space resort having room for approximately 100 guests by the year 2017 (TTG, 1999).

Summary

- Demand for tourism is not something that occurs by chance but is symptomatic of changes in the environment of societies that tourists originate from. Economic, social and cultural processes associated with the Industrial Revolution have had a major effect in shaping patterns of contemporary tourism.
- Tourism is a highly complicated amalgam of different parts. These include the environments of where people come from and where they go to. It is useful to think of tourism as a system, including a range of different inputs such as government policy, entrepreneurial activity, and human and natural resources. Critical to the system is the tourism industry which 'processes' these inputs to meet the needs of the tourists. Outputs of the system include economic opportunities and changes to the cultural and physical environments of destinations. All the time the system is subject to a range of external forces existing in society.
- There are now over 800 million international tourist arrivals per annum in the world and the number is expected to be over 1,600 million by the year 2020. Tourism is encompassing an increasing range

of physical environments and indigenous cultures as tourists demand
new experiences. In turn this means a growing number of
environments are being changed by tourism.

Further reading

Hobsbawn, E. (1962) *The Age of Revolution*, London: Abacus.

Towner, J. (1996) *An Historical Geography of Recreation and Tourism in the
Western World: 1540–1940*, Chichester: Wiley.

Urry, J. (1990) *The Tourist Gaze: Leisure and Travel in Contemporary Societies*,
London: Sage.

Suggested websites

United Nations World Tourism Organization: www.world-tourism.org

World Travel and Tourism Council: www.wttc.org

2 Perceptions of environments for tourism and ethical issues

- The meaning of environment
- Changing perceptions of the natural environment
- Tourist motivations and types of tourists
- Ethical considerations of tourism

Introduction

The environment is often referred to as the key component of tourism. Chapter 1 introduced the idea of viewing tourism as a system, linking the environments where tourists go to with the society they come from. This chapter examines the meaning of the term 'environment', how tourists perceive and interact with different environments, and raises ethical considerations of how the natural environment is used for tourism.

What is meant by 'environment'?

The state or well-being of the 'environment' has become along with global terrorism the most discussed contemporary global issue. The term is often, perhaps normally, equated with nature, and the basis of the debate rests upon the effects we are having upon it. However, the expression of 'environment' is also frequently used to refer to what actually surrounds us, e.g. our 'home' environments and 'destination' environments.

A useful classification of ways the environment may be defined is given by Attfield (2003) as:

1 *'the surroundings'*. This is the most common meaning and is associated with an individual's surroundings for the duration of their life or a society for the duration of its existence;
2 *'objective systems'* of nature, for example, mountains, rainforests, coral reefs, seas and rivers, which encompass society but precede and succeed it; and
3 *'perceived surroundings'* of an individual or animal that lend a sense of belonging and home.

It is evident from these interpretations that the environment can be understood as something that can be 'real' and with an independent existence of its own, or it can be understood as an interpretation and construct of the human mind. It is not the aim of this book to enter into an ontological debate about the environment, but it is important to be aware that there exists a philosophical debate about whether there exists a 'real' environment, or whether the environment is a construct of our own mind. Certainly *how* we perceive the environment and our place in relation to it will influence the types of values we attach to it, and be subsequently influential in determining how it is used. As Bruun and Kalland (1995:1) observe: 'In environmental studies it has commonly been assumed that there exists a fundamental connection between a society's management of natural resources and its perception of nature.'

It may seem strange even to consider how we perceive our surroundings, as we are very rarely encouraged to think in this way, perhaps assuming that our own society's constructs of the environment and nature are universally shared. However, different cultures often reveal distinct variations in how they view nature, supported by their spiritual and religious beliefs which influence our interaction. In the west, environmental problems became more evident in the 1960s. Lynn White Junior in the seminal paper 'The Historic Roots of Our Ecologic Crisis' attacked the Christian belief system as being the root cause of these problems. According to White, Christianity emphasised the dominion of man over nature and the use of the environment for his benefit and enjoyment.

Although White's work has subsequently been criticised for being too generalised and simplistic, it raised the issue of how our belief systems influenced our relationship with nature. Even within religions there may be contradictions; for example, within Christianity, there are two opposed traditions: the first is that humankind is unique in being made in the 'image of God and therefore has the right to behave in a god-like manner to the rest of the cosmos'; and the second is that 'humans are part of God's Creation just like the rocks and the trees and that no one part of this is inherently superior to another' (Simmons, 1993: 129). Consequently, whilst some Christians would support the former position of 'dominion' of humans over nature on the premise of being created in God's image, others would adopt the latter view of 'stewardship' of nature.

Most commentators take the view that the dominant paradigm in western society is supported by the first tradition, in which dominion over the earth is paramount, a projection of empire mentality on to nature (Hooker, 1992). As Singer (1993) points out, according to the Old Testament, God did drown almost every animal on earth apart from those that were led into Noah's Ark, to punish human beings for their wickedness. In the view of Nash (1989), Christianity is the most anthropocentric religion the world has seen, in which the earth is seen as a halfway house for trial and testing before death and hopeful admittance to heaven. Therefore in the Judaic-Christian religion, the earth does not embody God, instead removing God to on high. Singer (1993: 267) comments:

> According to the dominant Western tradition, the natural world exists for the benefit of human beings. God gave human beings dominion over the natural world, and God does not care how we treat it. Human beings are the only morally important members of this world. Nature itself is of no intrinsic value, and the destruction of plants and animals cannot be sinful, unless by this destruction we harm human beings.

In contrast, the relationship of many indigenous peoples to the land and animals incorporates them as part of the same environment, vis-à-vis being externally located to it. For example, in the case of Mayans, every human has an animal counterpart and every animal has a human counterpart, thus to harm one is simultaneously to harm the other (Hogan et al., 1998, cited in Whitt et al., 2003). Importantly in the context of establishing links with nature, the genealogies of indigenous peoples usually do not confine themselves to the human, including places and non-human beings (Whitt et al., 2003). For example, the Maori express their relationship to the land through 'whakapapa', identifying themselves by reference to mountain and river, and to their ancestral dwelling with the tribal landscape. Identification and transference of the human to the non-human results in entities of the landscape taking a spiritual form. This spirituality is expressed by the indigenous peoples of the Andes as *'allyu'*, a group of related beings living in a particular place, including the human and the non-human (Whitt *et al.*, 2003). Critically, this sense of interconnectedness means that the non-human world is not viewed as something external, but as an integral part of the people.

Indigenous belief systems that emphasise the interconnectedness of the human and non-human world are now influencing western thought in how we interact with nature. The raft of environmental problems that now face global society dictate that a new paradigm (or model) of how we interact with nature is demanded. For example, Chapter 6 discusses in detail the notion of sustainable development, which is often held to be a paradigm shift in western thinking about how we use natural resources for development. Based broadly upon the conservation of natural resources and a consideration of how our actions may harm or benefit future generations, it emphasises long-term thinking in decision-making. Its origins can be traced back to indigenous belief systems that existed before the advent of the agricultural and industrial revolutions. For example, in the Maori concept of 'Kaitiaki', it is emphasised that it is incorrect to think that we are guardians of nature, rather it is apt to think that the earth is a guardian of us. Perhaps even more pertinently, the paramount duty of First Nation Haudenosaunee chiefs is to ensure that their decision-making is guided by consideration and well-being of the 'seventh generation'. The well-being of the 'seventh generation' applies not only to the seventh generation in the future but also to the seventh generation in the past. Through this consideration, gratitude is expressed to the past seventh generation for what they sacrificed in order that natural resources were available for the present generation (Lyons, 1980).

Other religious belief systems also place humans in a position of being directly linked to nature. For example, the Chinese Daoist tradition emphasises a tripartite relationship between *Tian* (heaven), *Di* (Earth) and *Reiv* (Man), in which human beings should model themselves after *tian* and *di*, learning not to overvalue human concerns (Lai, 2003). Hinduism also has a long tradition of environmental protection under the concept of *ahimsa*, meaning non-injury. *Ahimsa* is associated with, though presupposes, the doctrine of 'karma' and rebirth, with the soul taking birth in different life forms such as birds, fish and animals, besides humans (Dwivedi, 2003). Thus there exists deep opposition within the Hindu religion to the institutionalised killing of birds and animals for human consumption. Similarly, in Buddhism, the taking of life is forbidden and vegetarianism advocated. In Islam, humans are stewards of the gifts of Allah, and this includes knowledge and understanding of environmental problems. Yet, despite these belief systems, economic development in countries like India and China has caused extensive environmental degradation as they attempt to meet the needs of their citizens and attain a western standard of living. As Simmons (1993: 133) points out:

> It has to be said that in both East and West many religious traditions have collaborated with human behaviour that is destructive of species and habitat, and with non-sustainable development. In the West, obviously, there has been little sieving of technology and much talk of the conquest of nature; in the East no guidelines have been elaborated for alternative forms of economic and social growth that are ecologically sustainable. In all, some reconstruction of the historic faiths seems to be needed if they are to contribute to an evolutionary *modus vivendi.*

THINK POINT

How would you explain the term environment? Do you feel connected to the environment or separated from it? How would you explain to someone the kinds of values you place on nature? How do you feel these values influence your behaviour towards nature?

Changing perceptions of landscape

Just as cultural and religious belief systems influence our interaction with the natural environment, they also shape what we regard as being 'beautiful'. Evidently, tourism would not exist without us having favourable perceptions of the natural and cultural attractions that exist in destination environments, as indicated in Figure 1.1. Yet, what we perceive as 'beautiful' landscapes and 'exotic' cultures are subject to modification as society and fashion changes. As Urry (1995) observes, a person's wish to visit a particular environment is something that is socially constructed, and depends upon developing a 'cultural desire' for a particular landscape. Shifts in perception of what are regarded as desirable landscapes are associated with social and cultural changes in society. For instance, a notable change in the itinerary of places to visit within the Grand Tour began in the mid-eighteenth century, as the desire to view picturesque and romantic landscapes became stronger (Towner, 1996). This marked a significant shift from what had previously been regarded as being desirable landscapes. The previous landscapes of fashion were those of the European low countries; that is, Belgium and Holland, because they illustrated the human ability to control and dominate nature to provide agriculturally productive terrain. This view of desirable nature as being agriculturally productive is representative of the separation of landscapes into 'controlled' and 'wild' areas, which according to Short (1991) developed approximately 10,000 years ago, with the onset of the agricultural revolution and a move to a more settled pattern of living. Previously in nomadic hunter and gatherer societies, no distinction was made between wilderness and the surrounding environment, as humans were too much part of nature to conceive of wilderness as being something separate from themselves.

Two main perspectives on the meaning of 'wilderness' can be recognised. The first is a 'classical perspective', in which the view is taken that the creation of liveable and usable spaces, such as urban areas, is a mark of civilisation and progress. The second approach is the 'romantic', in which untouched spaces have the greatest value, and wilderness assumes a deep spiritual significance. The shift to a desire to visit 'wildscape' in the mid-eighteenth century was marked by a preference for the raw power of nature, as manifested in mountains, gorges, waterfalls and forests. Until this time barren and mountainous landscapes had been largely detested and even feared by the majority of the population. For example, Smout (1990) points out that up to the eighteenth century the environment of the Scottish Highlands in Britain had been largely seen as an inhospitable place to think of visiting, and any use of its environment had strictly been in utilitarian terms, such as for the cutting of wood and mining of ores. Urry (1995: 213) adds: 'It is only in the last century [the nineteenth century] that a traveller passing through the Alps would have the carriage blinds lowered to make sure they were not unduly offended by the site.'

Urry (1990) associates the shift in emphasis towards wilder landscape with the development of the 'Romantic movement', which emphasised the feelings of emotion, joy, freedom and beauty that could be gained through visiting 'untamed' landscapes. The Romantics were a collective movement of European literary, artistic and musical figures, such as Rousseau, Coleridge, Wordsworth, Chopin, Goethe, Walter Scott, Hugo, Liszt and Brahms, who highlighted the importance of emotional experiences and having feelings about the natural and supernatural world. The movement was in part a reaction to the scientific thinking of the

Enlightenment period, and also to the growing urbanisation of Britain and western Europe, associated with the Industrial Revolution. Indeed, the Romantics represented a form of political opposition to what they perceived to be a loss of community, associated with the migration of people to urban areas from rural environments. One of the most poignant examples of the work of Romantic literature is the poem, 'The Daffodils', written by William Wordsworth at the beginning of the nineteenth century. The opening verse of the poem is well known:

> I wander'd lonely as a cloud
> That floats on high o'er vales and hills,
> When all at once I saw a crowd,
> A host of golden daffodils,
> Beside the lake, beneath the trees
> Fluttering and dancing in the breeze.

This verse emphasises the sense of freedom and wonderment about the physical environment that was an essential ingredient of the Romantic movement. It was written about an area known as the English Lake District, which is now a premier area for tourism in the United Kingdom, attracting millions of visitors per annum, many of them visiting Wordsworth's house near Grasmere. Ironically, Wordsworth was an opponent of mass tourism, sending a poem in 1844 to the prime minister condemning the building of a railway to the Lake District. Recently, his poem has been turned into a rap number and pop video in an attempt to attract more young people to the Lake District (Wainwright, 2007).

This change in desire, away from visiting landscapes reflecting human dominance to more wild landscapes, was a notable change in the western cultural perception of landscape, and one that has had a major effect on the patterns of contemporary tourism. For instance, Romanticism not only led to the appreciation of mountain areas, but also the coastline. A key icon of contemporary tourism, the beach, underwent a major perceptual change dating from the Romantic period. Just as mountain environments had predominantly been viewed as hostile and inhospitable places to visit in the eighteenth century, similarly the beach was also viewed with trepidation. For instance, in Daniel Defoe's early eighteenth-century novel *Robinson Crusoe*, the desert island was not equated with escapism and a type of 'paradise' as it is now, but viewed as an environment that was symbolic with abandonment and desolation. To the European explorers of the 'New World', beaches were points of initial

contact with foreign culture and potential battlegrounds, as Lĕcek and
Bosker (1998: xxi) comment:

> They were anxiety-ridden strips of no-man's-land where, according to
> the journals of Columbus, Corts, Cook, and Bougainville, Europeans
> first set eyes on others who, though like them, were yet utterly alien.
> Here, duels to the death were waged between races and cultures.

It is difficult to give an exact date of when the beach became a popular
place to visit for recreational purposes. Although its popularity is often
associated with its patronage by the middle and upper classes from the
eighteenth century for reasons of health and to escape urbanisation, Towner
(1996) notes that the coast already existed as a recreational space in folk
culture in the Baltic, North Sea and the Mediterranean, with a peasant sea
bathing culture existing in regions as diverse as the north of England and
the Basque region of Spain. However, the impact of urbanisation during the
eighteenth and nineteenth centuries was influential in encouraging a seaside
culture as Towner (1996:170) comments: 'Grafted onto the growing taste
for coastal areas was the influence of health awareness and desire to escape
from the effects of rapid urbanisation.' The poor social conditions of the
urban landscape encouraged the middle classes to develop anti-urban values
and helped make the beach more popular (Soane, 1993). The natural
environment of the beach offered a direct contrast to the squalid urban
conditions that existed in many urban areas, offering a sense of both
naturalness and timelessness. An example of the squalid conditions
experienced by many urban working-class people is given by Engels in his
description of life in nineteenth-century Manchester in England:

> In a rather deep hole, in a curve of the Medlock and surrounded on all
> four sides by tall factories and high embankments, covered with
> buildings, stand two groups of about two hundred cottages built
> chiefly back to back, in which live about four thousand human beings,
> most of them Irish. The cottages are old, dirty, and of the smallest
> sort, the streets uneven, fallen into ruts and in part without drains or
> pavement; masses of refuse, offal and sickening filth lie among
> standing pools in all directions; the atmosphere is poisoned by the
> effluvia from these, and laden and darkened by the smoke of a dozen
> tall factory chimneys.
>
> (Engels, 1845: 98)

In the latter part of the nineteenth century, the development of purpose-
built attractions also aided the evolution of seaside resorts into major
tourism destinations. The focus of beach tourism began to shift in this

period from health to pleasure, a change described by Urry (1990: 31): 'In the mid-nineteenth century this medicalised beach was replaced by a pleasure beach, which Shields characterises as a liminal zone, a built-in escape from the patterns and rhythms of everyday life.' In the first half of the nineteenth century, the ability to escape to the coasts for long periods of time was limited to the wealthiest of society and the development of coastal residential resorts offered an exceptional degree of privacy to the new elite, away from the industrial urban centres and the proletarian masses (Soane, 1993).

The emergence of a popular seaside culture was facilitated by the development of the railway network from the cities to the coast, which permitted a middle-and working-class holiday boom during the late nineteenth and early twentieth centuries. Villages and towns on the coastlines near industrial centres were transformed with promenades and piers, providing profits from previously economically redundant areas of cliffs and bays. The coast seemed to exercise an allure that eventually permeated all the social classes, representing a special place in many people's lives, referred to by Towner (1996: 212) as the 'geography of hope'. Certainly, the combination of a natural landscape consisting of sea, sky, cliffs and beach, with constructed features of promenades and piers, provided a distinctive sense of place away from the ordinary and the routine. Today the beach represents a major focus for recreational tourism, a 'liminal' zone of visitation, a place that fulfils a variety of needs. The popularity of the beach as the primary focus for recreation and tourism is supported by empirical research based upon 542 households in England and Wales, as discussed in Box 2.1.

Box 2.1

The popularity of the beach

In a survey of 542 households in England and Wales, Tunstall and Penning-Rowsell (1998) found that the beach was rated a more enjoyable place to visit than national parks, lakes, rivers, woods, museums, leisure centres or theme parks. People were found to be most attracted by those areas of coastline that were perceived as being undeveloped, possessing characteristics of natural settings such as dramatic views and scenery, and offering peace and quietness. The research also discovered that one of the most important functions of the beach was its ability to reconnect people with

Continued

Box 2.1 continued

their past, when visits were made in childhood and adolescence, thereby permitting an element of continuity and timelessness in a rapidly evolving and changing society. Tunstall and Penning-Rowsell liken this continuity to rereading a favourite book or the family photograph album. The beach also provided an environment for family interaction, with two generations of a family often visiting the beach, and was also found to act as a stress reliever. The ability to be in contact with nature through visiting the beach is emphasised in the following passage:

> Perhaps one of the most important elements in the beach experience is, therefore, the opportunity it gives for tactile close up contact with the natural physical world. . . . There are very few natural environments where children and even adults are allowed, indeed encouraged, to poke about, pick up, touch, shape and play with its physical material and the creatures it supports – crabs, shellfish and worms.

(Tunstall and Penning-Rowsell, 1998: 329)

The desire of the public for the beach to act as a point of contact with nature and the past is emphasised: 'Thus, the general view appears to be that the seaside should be like the seaside always was' (ibid.: 330).

Although the phrase 'like the seaside always was' will have different interpretations for different people, the empirical research supports the popularity of wildscape in society. Such research is also invaluable in giving guidance to planners and developers about the appropriateness of different types of coastal development for recreational use.

Source: Tunstall and Penning-Rowsell (1998)

Besides the popularisation of coastal areas in the nineteenth century, another landscape to be used for tourism in this period was mountain areas, particularly for the development of winter sports. Towards the end of the nineteenth century, skiing began to develop as an international activity in the European Alps. Resorts such as Davos and St Moritz were already familiar to wealthier travellers for the purpose of health tourism, based on the 'cold cure', which was a popular treatment. By the end of the nineteenth century, the popularity of the European Alps had increased with the expansion of the railroads into Alpine valleys (Barker, 1982). St Moritz in Switzerland, St Gervais in France and Bad Gastein and Bad Ischl in Austria were established as health spas by the end of the nineteenth century.

However, in the 1890s a new type of traveller appeared in the Alps more intent upon hedonism than recuperation, with winter sports, including ice-skating and skiing, becoming fashionable and popular. The mountains became increasingly popular with upper-class Victorians from the beginning of the nineteenth century, as an escape from the growing urbanised areas of the Industrial Revolution. By the beginning of the twentieth century their popularity had become such that D'Auvergne (1910: 289) was led to write: 'The tide of fashion has in fact been largely diverted of recent years from Nice and Cairo to these snow bound wildernesses.' An essential part of this fashion was the development of skiing, notably popularised by Sir Arthur Conan Doyle in his crossing by ski from Davos to St Moritz in Switzerland, and its subsequent reporting in fashionable journals. The Alps now became the playground for Europe's elite.

By the 1920s, a significant winter season in the Alps based upon destinations accessible by the railways had already developed. An indication of the growing importance of the winter market in the Alps is demonstrated by statistics from the Austrian Tyrol. In 1924, 14 per cent of the overnight tourist stays in the Tyrol occurred in the winter season, which by 1933 had increased to 44 per cent of the annual total of overnight stays (Barker, 1982). However, the development of skiing as a recreation was not confined to Europe. In Elyne's (1942) account of skiing in the Australian Alps, she says that the first skiing there took place in 1897. Good and Grenier (1994) give an earlier date, stating that the first ski club in the region was established in the 1860s gold rush. From these origins an industry has developed that now serves millions of skiers visiting different mountain areas around the world.

What was originally an elitist activity progressively become one of mass participation during the latter part of the twentieth century. Governments eager to aid regional economic development have supported the development of facilities to supply this demand. The economic potential of skiing in Alpine areas had already been noted by the beginning of the twentieth century. D'Auvergne (1910: 280) comments:

> Winter sport! the Swiss delightedly awakened to the commercial possibilities of snow and ice. The canton Grisons or Graubunden – the largest in Switzerland – was the first to find foreign gold beneath the snowdrifts. . . . Naturally the rest of Switzerland is on the alert and

eager to share the good fortune of the largest canton. . . . Chalets were transformed into hotels, brand new hotels were run up not always to the delight of the aesthetic traveller.

The post-war French government also acted upon the realisation that skiing could aid regional economic development. Lewis and Wild (1995) explain how the purpose-built ski resorts, or 'ski factories' as Lewis and Wild refer to them, because of their emphasis on accommodating large numbers of skiers and construction from glass, concrete and steel, were developed in the late 1950s to aid regional development in France. The demand of the emerging 'mass leisure class', and the opportunities to aid regional economical and social development, drove the transformation of mountain landscapes in the Alps. There has subsequently been created an economic dependence upon the ski industry and associated tourism. It is critical to the combating of seasonality, ensuring that employment and livelihoods can exist all year round, and not be purely reliant on the summer seasons. However, the ski industry is dependent upon the regularity and reliability of snowfall, the medium- to long-term reliability of which is under threat from climate change associated with global warming. This is likely to be especially problematic for low-altitude ski resorts, as is discussed more fully in Chapter 8.

Promoting the environment as an attraction

The changing perceptions of landscapes, combined with the changing social and economic conditions of the nineteenth century, presented opportunities for entrepreneurs to begin to promote images of the environment to the public to encourage them to travel. Notably, the development of the railways, combined with the influence of tour operators such as Thomas Cook and Sir Henry Lunn, meant that by the beginning of the twentieth century, images of 'wild' landscapes and foreign cultures had become central to attracting people to travel. Two examples of the use of wildscape to attract tourists are displayed in Figure 2.1 and Figure 2.2.

Although it is not possible to tell from the black and white image reproduced in this book, the original poster shown in Figure 2.1 uses only two colours, black and yellow, to demonstrate the austere image of the cliffs and sea. This image of Carrick-a-Rede on the coast of County Antrim in Ireland conveys a sense of wonderment, risk and adventure.
It also emphasises, through the use of the imagery of a barely discernible

Figure 2.1 *Poster – Midland Railway: Cook's excursion to Ireland*

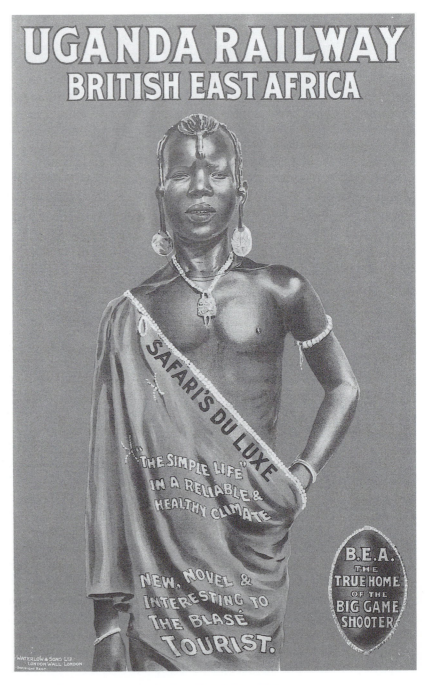

Figure 2.2 Poster – *Uganda Railway: British East Africa*

single figure on the suspension bridge, the power of nature and the solitary experience wilderness can provide. A further interesting aspect of the poster which dates to 1904, is that it was produced for the Midland Railways in association with Thomas Cook the tour operator, emphasising the role of both the railways and tour operators in developing tourism in this period.

In Figure 2.2, the image on the poster dating to 1908 presents a more 'exotic' kind of travel to the western eye. Again the poster demonstrates the importance of the railway companies in promoting and facilitating travel at this time. It also underlines the post-colonial nature of contemporary tourism, with the countries of Kenya and Tanzania which formed the then British East Africa, now representing two of the biggest receiving countries of British outbound tourists in Africa. The poster also raises ethical issues about tourism and the 'use' of indigenous cultures by the tourism industry for promotional purposes. Clearly, the exotic figure is presented to appeal, as in the wordage on the poster, 'new, novel and interesting to the blasé tourist'. The notion of a 'blasé tourist' at the beginning of the twentieth century may seem surprising, but it must be remembered that this kind of holiday was targeted at an elite in British society, hence the emphasis on 'safari's du luxe'. For this travelling elite, perhaps already over-familiar with tourism to Europe, new environments were being sought. The other wordage on the poster is also revealing, 'The simple life in a reliable and healthy climate', emphasising the desire for escapism from what was becoming an increasingly urbanised and complex environment.

However, the biggest attraction of travel to British East Africa at this time, as promoted in the poster, was the activity of big game shooting. The growth in the popularity of hunting in East Africa is associated with the arrival of white settlers in the nineteenth century. In Europe, hunting was an activity exclusive to the aristocracy, helping to differentiate them from other classes. The settlers now had an enormous shooting estate of their own and the guns proclaimed them as the new aristocrats. Within a few decades of the Europeans arriving, the blaubok and quagga were wiped out, both of whom had survived three million years of contact with the indigenous people, and men boasted of killing 200 elephants on safari. One notable example of the carnage caused through hunting was an expedition led by Theodore Roosevelt and his son, in which 5,000 animals of 70 different species were killed, including nine of East Africa's remaining white rhinos (Monbiot, 1995).

Although the imagery used by today's tour operators is perhaps more subtle than that in Figure 2.2, the messages conveyed in destination advertising continue to rely upon the differentiation of the destination environment from the one at home. This is aptly demonstrated in Figure 2.3, showing a typical image used by tour operators to sell long-haul travel to destinations outside Europe and America. This particular image is of one of the coral islands of the Republic of Kiribati in the Pacific, the first land area to welcome in the new millennium, and ironically one of the areas most under threat in this century from increasing sea levels resulting from global warming, a phenomenon discussed fully in Chapter 8. The physical aspects of the coral island, such as the palm trees and white beach, combined with the crystal-clear sea, are often utilised in promotional images by the tourism industry to attract western tourists.

The influence of the Romantic movement on contemporary copywriting in tourist promotion is evident in the following advertisement for the Azores. Under the title 'Europe's Last Frontier', the Azores are described as:

There's a place where beauty is not near extinction.
Where nature and man have found perfect harmony. Where peace is unbroken.
There's a place where you may still watch whales swimming freely in their natural habitat. Where vulcanic heat allows for an exquisite and tasteful way of cooking.
And where memories of Atlantis still linger along the quiet paths
and the evergreen tracks of the highlands.

Figure 2.3 *Image of Kiribati*

Underneath these words is an image of a whale with its tale protruding from the water. It is not a coincidence that the outline shape of the layout of the words of the advertisement resembles the tail of a whale. These words also place a heavy emphasis upon the nostalgia of the way the natural environment is often perceived as having been, in which man and nature are in harmony and symbiosis, a world that has changed little since the times of Atlantis.

Empirical research also supports the importance of 'unspoilt' environments being attractive for tourism. During market research with German tourists, when asked about the characteristics of a destination that give it 'quality', the majority of the statements with the highest percentage response rates were found to be environmentally linked, as shown in Box 2.2.

Box 2.2

The main characteristics for 'quality' tourism

Statement	Percentage importance rating
The landscape must be beautiful	46
The atmosphere must be relaxed	46
Cleanliness is a matter-of-course	39
The sun must shine	38
The climate must be healthy	32
Good cuisine must play a part	30
Quietness and very little traffic	29
The surroundings must be typical for the country	28
There must be attractive places for excursions	26

Note: The same person could name several characteristics if they thought them important in determining the quality of a destination.

Source: European Tourism Analysis (1993)

These results indicate that the perception of a 'quality' destination is closely associated with factors relating to its natural characteristics. Critically, in the eyes of the tourist, the landscape must be beautiful, a perception which as previously discussed is culturally determined and thus subject to change. The recuperative aspect of tourism is also evident, with emphasis being placed on a relaxed atmosphere, healthy climate, cleanliness and quietness. Cultural aspects of the environment are also stressed, with a wish for good cuisine and surroundings typical of the country. However, whether tourists would wish to be surrounded with the type of poverty seen in some regions of developing countries is debatable. The probability is that the majority of tourists would prefer surroundings that reflect their own image of a destination in a positive way, rather than being faced with the unpleasant reality of some aspects of the local society.

From the evidence of tourism advertising, it is suggested that images that suggest largely 'unspoilt' physical and cultural environments are essential for attracting tourists from developed countries. The way the tourist gains access to sample these environments is typically by purchasing a 'right' of access via the market system. In this sense, tourism can be interpreted as a form of consumerism, centred upon the consumption of experiences provided by foreign environments.

THINK POINT

Think about images of environments that are attractive to you. What makes them attractive? What kinds of associations do you make with them? How are they different to the environment in which you normally live?

Tourism as a form of consumption

As societies in the West have entered advanced stages of capitalism, 'consumption' has become a central theme of lifestyle. The concept of consumption surpasses the idea of merely buying something to meet basic physiological needs, such as buying a loaf of bread because of hunger. Using social and human psychological perspectives associated with consumer behaviour studies, it is recognised that by purchasing a certain good or service, a range of needs may be met that go beyond our most basic biological requirements. These include social needs (such as feeling part of a group or having a sense of belonging), the need for love, the need for self- and social esteem, and the need for self-development and self-fulfilment. Such ideas are linked closely to the work of Abraham Maslow (1954) in the field of human psychology.

Sociologists emphasise that through the process of consumption, it is possible to differentiate oneself from the crowd, and subsequently gain a sense of identity. The idea of identity being linked to consumption was a theme developed in the late nineteenth century by the American sociologist Thorstein Veblen (1899). Veblen's work was based upon observations of the newly emerging wealthy middle class in America, who were making considerable amounts of money from trade and manufacturing. In what was a critical account of their lifestyle, Veblen coined the phrase 'conspicuous consumption', to explain how this class used leisure and consumption to differentiate themselves from the rest of society. Through the purchasing of goods and other ornaments and services, they took on a different identity from the rest of the populace, who could not economically afford to purchase these products. By the end of the nineteenth century, tourism also offered a form of consumption, which permitted social differentiation. Within what was a predominantly patriarchal society, by sending one's wife or daughter on holiday from America to Europe, the message was conveyed that one possessed wealth.

At a similar time to Veblen's observations of what he termed the new 'leisure class' in America, in Germany another sociologist, Georg Simmel, was also observing and writing about the life of Berliners. His central theme was based upon the interaction between the individual consciousness and the modern city. According to Simmel, the complexity of urban living, combined with the need to filter out stimulation to a manageable level, leads the individual to an ultimate sense of indifference about the city, whereas at first it had been perceived as exciting and stimulating (Lechte, 1994). Simmel also emphasised the need for an individual to preserve a level of autonomy in urban areas (Bocock, 1993). One way of doing this, as observed by Veblen, is through 'conspicuous consumption'. In this sense it can be said that the process of urbanisation had a major influence in encouraging consumption beyond meeting basic needs, to provide a form of identity.

Using tourism as a means to achieve social differentiation has become increasingly prevalent in western society. Mowforth and Munt (1998) use the term 'habitus', borrowed from the French sociologist Pierre Bourdieu, to suggest that the types of tourism we participate in carry cultural symbols and meanings. Habitus refers to the ability and inclination of individuals and social classes to adopt objects and practices that differentiate them from others in society. Although the ability to participate in international tourism in contemporary society fails to achieve social differentiation as at the time of Veblen, yet where we are able to go and what we choose to do on holiday carries cultural messages and differentiates us from others in society.

The link between urbanisation, consumption and tourism has been developed by both Boorstin (1992) and MacCannell (1976) to explain why people choose to travel. Boorstin proposed that tourism represented nothing more than a form of escapism from urban life, serving little purpose beyond mindless enjoyment. Central to this idea is that the tension of urban living creates a state of anomie in the individual. The concept of anomie was developed by the sociologist Émile Durkheim, to describe the individual's feeling of isolation in a society where the norms governing interaction have broken down, and society has transcended into a state of lawlessness and meaninglessness (Dann, 1977; Slattery, 1991). Viewing tourism as a commodity and questioning its degree of meaningfulness, Boorstin (1992: 91) makes the following comments upon a visit to Hyderabad in India:

> Seated beside me on the plane were a tired, elderly American and his wife. He was a real estate broker from Brooklyn. I asked him what was interesting about Hyderabad. He had not the slightest notion. He and his wife were going there because the place was 'in the package'. Their tour agent had guaranteed to include only places that were 'world famous', and so it must be.

In contrast, MacCannell (1976) views tourism not as a function of escapism, but as a pilgrimage of modern man, a quest for authenticity which is felt to be lacking in post-modern societies. MacCannell adds that the desire to get in touch with a way of life that has disappeared with the growth of urban metropolises, also helps to confirm the relative comfort and material well-being of the tourist, vis-à-vis the 'primitive'. For example, in MacCannell's latter argument, the purpose of a visit to British East Africa, as portrayed in Figure 2.2, would partly be for the tourist to confirm their way of life as being superior to that of the life of the indigenous person. Although the early works of Boorstin and MacCannell are important for providing early sociological perspectives on tourism, they present two extreme views of the influence of society on motivating people to become tourists. Subsequently, Cohen (1979) criticised their work as being overstated, and suggested there was a lack of sufficient empirical evidence to support either claim. It is perhaps best to picture a continuum between the polarised positions of Boorstin and MacCannell, along which the influence exerted by society upon the individual to become a tourist, rests somewhere in between. However, empirical evidence to support the proposition that urban areas necessarily increase participation in tourism is difficult to find, mainly because to date there seems to have been judged little necessity to investigate this relationship. Although slightly historical, the results of one survey that supports the observation that urbanisation encourages tourism are shown in Box 2.3.

At face value, the positive correlation that exists between the increasing size of the urban area and the increasing departure rate for summer holidays lends support to the theory that urban life encourages tourism, whether for escapism or as a quest for something more meaningful. However, as Tourism Planning and Research Associates (1995) point out, other factors favour the participation of urban dwellers over rural dwellers in tourism. These include higher incomes, smaller households with a corresponding greater flexibility for travel, paid holiday leave, and a freedom from being tied to elements of the agricultural cycle such as looking after livestock.

Box 2.3

Percentage of population taking a summer vacation in France by settlement type, 1969–88

Settlement type	1969	1985	1986	1987	1988
Rural	19	40	38	40	43
Less than 20,000 inhabitants	38	51	46	47	47
20,000 to 100,000 inhabitants	51	60	53	55	60
100,000 + inhabitants (excluding Paris)	56	64	62	60	60
Paris	83	83	79	75	79

Source: Tourism Planning and Research Associates (TPR)(1995)

Although the theories of Boorstin (1992) and MacCannell (1976) are useful in initiating the debate about tourism motivation, subsequent work in the fields of social psychology and sociology suggests that tourism motivation is a highly complex phenomenon, satisfying a variety of individual needs. This complexity means that there is a lack of an agreed psychological theory to underpin an approach to understanding tourism motivation. As Philip Pearce (1993: 114) comments: 'research into why individuals travel has been hampered by the lack of a universally agreed upon conceptualisation of the tourist motivation construct.'

In one of the most significant empirical studies of tourist motivation based upon tourists staying in Barbados, Dann (1977) found that escapism and the search for social status formed the two most important reasons to participate in tourism, which he subsequently termed 'anomie' and 'ego-enhancement'. Behaviour associated with a feeling of anomie manifested itself by tourists placing an emphasis upon social interaction with family and friends. The need for ego-enhancement is, according to Dann, a reflection of people being denied status in their home society, leading to a desire for a higher status away from the home environment. By visiting a country where the standard of living is substantially lower than in the west, a tourist can live in a style of comparative luxury, albeit temporarily. Furthermore, by indulging in a type of lifestyle that cannot be enjoyed by the majority of local people, the tourist is conferring their status upon others. Additionally, according to the concept of 'habitus', by visiting the 'right' type of destination, it is possible to impress one's peer group, thereby raising one's status.

Besides the needs of social bonding and social esteem, other kinds of psychological needs are also important in motivating people to become tourists. Other suggested psychological determinants of tourism are shown in Box 2.4 below.

Empirical research lends support to tourism fulfilling a range of psychological needs. The desire for a change of lifestyle and the stimulus of a new environment is emphasised by Jamrozy and Uysal (1994), who conducted an analysis of the motivations of the German tourist market. They found that escape and a search for novelty and experience were the two major driving forces. Other motivations included: being with friends and togetherness; adventure and excitement; doing nothing; and prestige. The multivariate nature of the needs that are satisfied through tourism is a theme that has been developed by Pearce (1988), who suggests that

Box 2.4

Psychological determinants of demand

- Escape from a mundane environment, akin to Boorstin's (1961) theme of escapism
- Relaxation, including mental and physical recuperation
- Play, a chance to regress to a childhood state which is not available to adults within the constructs of everyday society
- Strengthening family bonds
- Prestige – akin to Dann's (1977) idea of ego-enhancement
- Social interaction
- Sexual opportunity
- Educational opportunity
- Self-fulfilment – akin to MacCannell's (1976) idea of self-discovery.

Source: After Ryan (1991)

people have 'careers' in tourism, just as at work. As in a work career, a travel career is both consciously determined and purposeful, although an important difference is that a career in tourism is more likely to be intrinsically motivated than a work career, which for the majority of people is likely to be extrinsically motivated. The basis of the 'travel career' is that motivations for participating in tourism are dynamic and will change with age, life cycle stages and the influence of other people. Additionally, past tourism experiences are likely to influence future behaviour.

In summary, the reasons why people choose to travel are complex and multivariate. The influence of one's home society would seem to be critical in shaping the desire to travel but there is no single explanation of why people choose to travel. However, what is evident from the increasing demand for tourism is that growing numbers of people identify tourism as a medium through which to satisfy their wants and desires.

> **THINK POINT**
>
> What were your motivations for taking your last holiday? To what extent do you feel your motivations were influenced by your home environment, including both people and your physical surroundings?

Types of tourists

Given the complexity of what motivates people to become tourists, it is unsurprising that tourists choose to visit different types of destinations and display various types of behaviour within the destination environment. Differences in behaviour are compounded by a range of interrelated factors, such as demographics, culture, life cycle, level of education, and beliefs and attitudes. The fact that tourists are different means that their interaction with the cultural and physical environments of the destinations they visit will vary. The realisation that tourists are not all the same has meant that various attempts to classify tourists into different types have been made. One of the most acknowledged phenomenological attempts to classify tourists' experiences was made by Cohen (1979). He identified five main categories of tourists, as shown in Box 2.5.

Cohen's (1979) central theory to explain the different types of tourist experience is that the significance of tourism depends upon the individual's 'total world-view'. This means whether the person orientates to a 'centre', which gives spiritual or cultural meaning to their life, and if so, where this

---

Box 2.5

A phenomenology of tourist experiences

Mode	Behaviour
Recreational	Emphasis is placed on tourism as a form of entertainment, restoring physical and mental powers, and endowing a sense of well-being. The tourist is thereby refreshed to return to their society where their centre lies.
Diversionary	The tourist finds in tourism a diversion or escape from the boredom and meaningless of everyday life. This type of person has no centre but neither are they looking for one.
The Experiential	The tourist looks for meaning of life in the culture of the 'other', as the tourist has lost their own centre in their home society. Whilst the tourist observes and interacts with other cultures, she/he remains aware of their 'otherness'.

Box 2.5 continued

The Experimental	This surpasses the experiential mode in the degree of experience of foreign culture. The tourist experiments in different ways of living, e.g. on a kibbutz, in search of a spiritual centre. However, they remain unsatisfied by the authenticity of any of the cultures they have visited.
The Existential	The individual's spiritual centre is now located in another place away from the home environment. For practical reasons, such as work or the family, the person may not be able to relocate physically but will visit whenever possible.

Source: After Cohen (1979)

'centre' is located in relation to the society in which they live.

The behaviour of a tourist in a destination will therefore be related to where their centre lies. For example, the 'recreational' tourist would be primarily concerned with rest (or 're-creation') and enjoyment rather than seeking a spiritual centre away from home. Interest in and contact with the local culture is likely to be kept to a minimum and this type of tourist is likely to be content with 'pseudo-events' and inauthenticity. They may choose to surround themselves with the paraphernalia of home, such as familiar daily newspapers, food and drink, emphasising a close contact with the home environment. An example of the development of local tourism services to satisfy this kind of market is shown in the photograph (Figure 2.4) of a local bar in Lanzarote. In this case the English tourist can feel safe, that upon entering 'paradise', they will find an English manager.

In the 'diversionary' mode, tourism acts as a diversion from the boredom and meaninglessness of everyday life, a state of alienation that is experienced in one's home society. Unlike recreational tourism, the individual's participation in tourism does not re-establish adherence to a meaningful centre in their home environment. Nor is this type of person looking to be centred in another culture, rather they are centre-less.

A tourist having an 'experiential' experience is involved in the process of searching for a centre away from their home environment. The behaviour of these tourists will reflect a much higher degree of contact with the local cultures than the previous two types, involving an exploration of

Figure 2.4 *'Paradise' is placed under English management*

their ideas and value systems. However, the experience-orientated tourist remains aware of the 'otherness' of the culture they are staying in, as Cohen (1979: 188) comments: 'the experience-oriented tourist, even if he observes the authentic life of others, remains aware of their "otherness", which persists even after his visit; he is not "converted" to their life, nor does he accept their authentic lifeways.'

The 'experimental' mode involves trying to live in different cultures with the aim of discovering a feeling of authenticity and a new centre. This type of experience is likely to have a higher level of involvement and immersion with the local culture than for the experience-oriented tourist. However, despite experimenting with different cultures, the tourist remains unconvinced by the authenticity of any particular one, and carries on searching. In the 'existential' mode, the tourist has relocated their centre to another culture, but for practical reasons, such as work and family commitments, cannot physically relocate to their elected centre. They would therefore visit their elected environment as often and for as long as is possible within the constraints of everyday life.

Other typologies based upon tourists' behaviour have also been developed. In an earlier categorisation, Cohen (1972) recognised four main types of tourists, based upon the relationship between tourists, the destination environment, and tourism businesses. These are the 'organised mass tourist', the 'individual mass tourist', the 'explorer' and the 'drifter', as shown in Box 2.6.

Cohen categorised the first two typologies, 'the organised mass tourist' and 'the individual mass tourist', as being 'institutionalised' tourists.

Box 2.6

Cohen's (1972) typology of tourists

Category	Characteristics
The organised mass tourist	Highly organised travel; minimum contact with destination culture; travel in large groups.
The individual mass tourist	Rely on the tour operator to arrange flights and accommodation; enjoy an element of liberty but will still tend to stay on the 'beaten track'.
The explorer	Tries to avoid the tourist track; make their own travel arrangements; learn the language of the place they are going to and attempt to associate with local people; retain some of the values and routines of home life.
The drifter	Attempts to become part of the local community by living and working with them; shuns contact with other tourists and the tourism industry.

The other two types, 'the explorer' and 'the drifter', he categorised as 'non-institutionalised' tourists. By institutionalised, Cohen implies a heavy dependence upon the tourism industry to organise their holiday. In contrast, non-institutionalised tourists place little reliance upon the tourism suppliers, searching for authentic experiences with an itinerary they choose.

Other researchers, notably Plog (1974), have tried to link personality characteristics to particular types of tourist. Plog's work is based upon the concept of psychographics, which has its origins in psychoanalysis. Its application to tourism involves examining and trying to comprehend the tourist's intrinsic desires to choose a particular type of holiday or destination. Plog's seminal work was developed in a commercial consultancy setting, working with airline business clients in the USA. Concerned with understanding the personality characteristics of people who were flyers as opposed to non-flyers, Plog used telephone interviews to ask questions relating to personality characteristics. He subsequently established a continuum of typologies ranging from what he termed 'psychocentrics' to 'allocentrics'. The psychocentrics were found to display characteristics of being self-inhibited, nervous and non-adventuresome, whilst allocentrics showed traits of being variously experimental, adventurous and confident. According to Plog (1974), the population rests upon a continuum ranging from allocentrism to psychocentrism, along which it is normally distributed.

These personality characteristics were also found to influence the type of destination that would be sought by the tourist, as shown in Box 2.7.

Allocentrics are likely to enjoy the sense of discovery and new experiences of destination environments before other tourists arrive there. It should be stressed, however, that having an allocentric personality does not automatically imply a respect for the local environment. For instance, white-water rafting on an inaccessible river in the Himalayas would be a type of holiday matching the preferences of allocentrics, but does not necessarily imply a respect for the local cultural or physical environment.

Plog (1974) also linked the typologies to stages of destination development. As allocentrics tell their friends about their recent travels, these destinations become the latest fashionable or 'in' spots. The numbers of tourists increase, hotels and other tourist facilities are developed, and the destination orientates towards serving the mid-centric market. Tour operators are now likely to be bringing in groups of tourists to the destination. The allocentrics now begin to leave the destination as

---

Box 2.7

The characteristics of psychocentrics and allocentrics

Psychocentrics

- Self-inhibited, nervous, non-adventuresome
- Search for familiarity and symbols of home, for example, newspapers, drink and food, and an absence of foreign atmosphere
- Favour institutionalised tourism
- Prefer a high level of tourism development

Allocentrics

- Experimental, adventuresome, confident
- Prefer novel and different destinations and enjoy experiencing new environments
- Prefer non-institutionalised tourism allowing freedom and flexibility
- Prefer non-chain-type accommodation and few purpose-built tourist attractions

Source: After Plog (1974)

---

it has lost the characteristics of novelty and being unspoilt which attracted them there initially. At this mid-point in the cycle of development, the destination is attracting the maximum number of tourists possible. If the destination continues to expand, then it risks the problem of environmental degradation and ultimately a reliance on artificial or constructed attractions to attract tourists offering 'sun and fun'. According to Plog (1974), such destinations are ultimately likely to appeal only to the small psychocentric market.

Besides the work of Cohen and Plog, numerous other typologies of tourists have also been created over the last 20 years. For instance, Dalen (1989), cited in Swarbrooke and Horner (1999), produced a four-group classification of Norwegian tourists. These included: (i) the 'modern materialists' whose main motivation is hedonism and gaining a tan to impress people when they get home; (ii) 'modern idealists' who also like partying but are more intellectual than the previous group and want to avoid mass tourism and itineraries; (iii) 'traditional idealists' who demand culture, peace, heritage and security; and (iv) the 'traditional materialists', who always look for special offers and low prices and have a strong concern with personal security.

Gallup (1989) also produced a typology of travellers for the American Express organisation. They identified the following: 'adventurers' who are motivated to seek new experiences from activities and other cultures and for whom travel plays a central role in their life; 'worriers' who display a considerable amount of anxiety about travelling and subsequently travel is not particularly important to them; 'dreamers' who are fascinated by travel but tend to orientate their trips to relaxation rather than anything too adventurous; 'economisers' for whom travel is not a central part of their life but travel because they need a break whilst seeking value for money; and finally 'indulgers' who like to be pampered and are willing to pay for a higher level of service when they travel.

Poon (1993) suggests that by the beginning of the 1990s, a new type of tourism consumer had emerged. Changes in society, including consumer sophistication, heightened levels of environmental awareness, and increased familiarisation with travel, have led to the emergence of what Poon terms 'new tourists'. She comments: 'New tourists are fundamentally different from the old. They are more experienced, more "green", more flexible, more independent, more quality conscious and "harder to please" than ever before.' Poon also defines new tourists as being spontaneous and unpredictable, and not as homogeneous as the old tourists. They are hybrid, reflecting changing demographic patterns and lifestyles in society, as Poon (1993: 9) observes: 'Families, single-parent households, people whose children have left home, "gourmet" babies, couples with double income and no kids (DINKS), young urban upwardly mobile professionals (YUPPIES) and modern introverted luxury keepers (MILKIES) are examples of lifestyle segments.' Over a decade after Poon's observations, independent travel is now an established part of the market. Major factors encouraging this trend include the evolution, success and expansion of low-cost (budget) airlines, and the widespread use of the internet by both the tourism industry and travellers.

However, although typologies are useful in alerting us to the point that tourists are not homogeneous, there are a number of criticisms that have been made of the attempts to produce typologies in the field of tourism research. Sharpley (1994) criticises typologies as follows: for being static and not taking account of variations in tourist behaviour or experience over time; for being isolated from the realities of the wider social setting, for example, financial reasons may force a young family to take a package holiday although the parents may have allocentric tendencies; and for the lack of common agreement amongst tourism academics on the number of tourism typologies and how to clarify their boundaries which leads to repetition and a confusion of terminology.

Experiencing the environment

The various typologies of tourists suggest that tourists are not homogeneous and are likely to be searching for different kinds of experiences from the destination environment. Although there is a marked absence of research into how tourists experience the environment, Ittleson *et al.* (1976) drew up a taxonomy of environmental experiences that was later adapted by Iso-Ahola (1980) to the field of recreation. These modes of environmental experience have been expanded to the field of tourism, to include a behavioural dimension, as shown in Box 2.8.

Box 2.8

Modes of experience of the destination environment

Mode of experience	Interpretation	Behaviour and environmental attitudes
Environment as a 'setting for action'	The environment is primarily interpreted in a functional way as a place for hedonism, relaxation and recuperation. The physical environment may also possess the characteristics necessary for the pursuit of activities, e.g. rivers for rafting, snow for downhill skiing, coral for scuba-diving. The pursuance of satisfying the needs of relaxation or excitement and thrills is paramount over environmental appreciation. The destination environment is primarily seen as external to one's self.	Conscious or subconscious disregard for the environment and a lack of interest in learning more about its natural or cultural history. In some cases a possible disinterest and disregard for environmental codes of behaviour, leading to negative environmental consequences. Examples would include littering, breaking of coral, frightening of animals, disregard for local customs and traditions.

Continued

Box 2.8 continued

Mode of experience	Interpretation	Behaviour and environmental attitudes
Environment as a social system	The environment is seen primarily as a place to interact with friends and family.	Physical setting becomes irrelevant as the focus of the experience centres on social relationships.
Environment as emotional territory	Strong emotional feelings associated with or invoked by the environment which provide a sense of well-being. The surrounding environment is now an important part of the tourist experience in terms of personal development and is capable of producing deep-felt emotions.	Sense of well-being and wonder at being in a different environment. The tourist may be moved to paint the environment, or write poetry about it, or sit or walk in contemplative wonder.
Environment as self	Merging of the physical and cultural environment with self. The environment ceases to be detachable from the person or external to them. The person's spiritual centre is now firmly located in this environment. Any damage or harm to the environment is perceived as damage to oneself. The destination environment may be experienced in an 'existential' way if physical constraints, e.g. work or family commitments, restrict the opportunity for the person to relocate there permanently.	Strong attachment to landscape(s) and culture(s) that are perceived as being 'better' than the home society. The tourist will have read extensively about their adopted cultural and physical environment and if necessary attempted to learn the language to allow them to interact with the local community in as meaningful manner as is possible.

Source: Developed from Ittleson et. al. (1976); Iso-Ahola (1980)

This interaction demonstrates a gradation starting from little interest in the environment, beyond its providing the setting to indulge in a certain type of behaviour, to one of much greater interest and attachment. As with all taxonomies of experience, the boundaries are not fixed and impermeable but are a reflection of the attitudes that a tourist may possess about the environment they are visiting. They are not exclusive and it is possible that an individual may experience more than one mode of interaction when visiting a destination.

It is likely that the attitudes of the tourist to the environment will be reflected in their behaviour. Tourists are likely to choose a destination environment because they feel it will be appropriate for the type of experience they are searching for. This means that the way environments are advertised carries implications for the behaviour of tourists and their subsequent impacts. For example, if environments such as mountain areas or coral reefs are advertised for activity-based kinds of tourism or as backdrops to high times and parties, then the possibility of a negative interaction between tourism and the environment is heightened. This scenario is exemplified in the following extract of an advertisement for a holiday in Tenerife from the brochure: *Summer '99: Have it your way*:

> We all piled into Bobby's, the biggest club in Americas, where they had this brilliant English band and a rather (ahem) adult comedian – certainly broadened our knowledge if not our culture. The reps were all there too and really getting into it. Fuelled by that, we exchanged girls with hip strings for whales with blow holes on Atlantic Knights. What a mad day . . . a Freestyle cruise out to sea to spot dolphins and whales and other aquatic things. Oh and to drink a large amount of free beer.
>
> (Club Freestyle, 1999: 32)

Although partying in the vicinity of whales and dolphins may not necessarily seem a particularly harmful activity, there is concern that too much human activity and noise stresses them. In New Zealand, Canada, South Africa and the United States, swimming with whales has been banned. There are also initiatives being taken by organisations such as the International Fund for Animal Welfare to control whale-watching, with limits on the noise and the number of boats, and restrictions on their proximity to animals (Harrison, 1998).

Similarly, if local cultures of destinations are displayed in tourist brochures in a way that relies upon stereotyping or shows a lack of respect, then the opportunities for cultural misunderstanding and conflict will be enhanced. For instance, in the copywriting of one tourism brochure in the early 1990s, Pattaya in Thailand was referred to thus:

> We'd guess there's getting on for a thousand bars in Pattaya . . . and then there's the nightclubs, discos, and restaurants. If you could imagine global TV announcing the end of the world in 24 hrs time, then what goes on in Pattaya all the time is what the approach to doomsday would probably be like. If you can suck it, use it, feel it, taste it, abuse it or see it, then it's available in this resort that truly never sleeps.

Given that tourists display different behaviour patterns, it is important for managing the environmental consequences of tourism that consideration is given in marketing strategies to attracting a type of tourist whose behaviour is likely to be compatible with the destination surroundings. Subsequently, it is possible that an understanding of environmental attitudes and ethics of markets is likely to have an important role to play in future destination marketing strategies, just as socio-demographic characteristics have had in the past.

How the environment is used for tourism: an ethical approach

Environmental philosophy and ethics

The issues raised in this chapter concerning our interaction with our surroundings, and the types of experiences that are sought in nature, raise questions about the type of relationship we have with the environment. This questioning of our relationship with the natural environment has been encapsulated in the philosophical genre of environmental ethics. Such ethical concerns have a resonance in the context of tourism, given it is predominantly based upon the consumption of experiences through an interaction with environments composed of wildlife, nature and indigenous cultures. As Krippendorf (1987: 20) puts it: 'The countryside, the most beautiful landscapes and the most interesting cultures around the globe form the theatre of operations of this industry.' Subsequently, tourism's interaction with these physical and cultural environments raises ethical questions of how they are used by the tourism industry and tourists. This ethical questioning of tourism is reflective of a broader concern over our interaction with the environment, which has arisen from a growing awareness of environmental problems resulting from human actions, as is discussed in the next chapter.

Although we may think of environmental concern as a relatively recent occurrence, in the nineteenth century, Henry Thoreau was already expressing reservations over the effects of development upon nature. Living at a time when America was rapidly industrialising, Thoreau observed changes in his home town of Concord, Massachusetts, including the arrival of the railway, the disappearance of familiar animal species such as the beaver and deer because of over-hunting, and the removal for use as fuel of much of the local forest he grew up with. He was highly critical of the way that capitalism encroached upon previously open land, enclosed it and fenced it off against trespassers. Written during his two-year stay in comparative isolation at 'Walden Pond', a glacial lake close to Concord, his most famous work *Walden,* published in 1854, questions how nature is used for capitalism. The work was influential in recognising a deeper value to nature than purely the instrumental and was influential in the later founding of the American environmental movement.

In a similar vein to Thoreau, though writing nearly a century later, Aldo Leopold also stressed the requirement for a more integrated and holistic relationship between humans and nature. Advocating the need for a 'land ethic', in his most famous work *A Sand Country Almanac*, Leopold (1949: 219) comments: 'In short, a land ethic changes the role of Homo Sapiens from conqueror of the land-community to plain member and citizen of it. It implies respect for his fellow-members, and also respect for the community as such.' Leopold was suspicious of recreation and tourism, viewing it as already having a negative impact on wildlife, and the travel trade as encouraging this trend. Besides their impact upon nature, he was particularly concerned about how the tools of advertising and promotion in the 1940s were being used to inspire access to nature in bulk, and consequently reduce the opportunities for solitude (Hollinshead, 1990).

Later in the twentieth century, as environmental problems began to take a higher precedence and the conservation of nature emerged as a serious consideration, the Norwegian environmentalist Aerne Naess (1973) identified two broad philosophical approaches. 'Shallow ecology' is based upon an 'anthropocentric' view of nature, meaning that nature is viewed as being separate from humans, and its value rests purely in terms of the use it has in meeting human needs and desires. Consequently, the anthropocentric view of why the environment should be conserved or treated in a responsible way, rests solely with the benefits this would bring for humans. By contrast 'deep ecology' rejects any separation of nature and humanity, stressing their interconnectivity, and that all beings are of equal value. A value is given to nature, which emphasises its right to existence, rather than its instrumental value. Thus rather than assuming that society should utilise natural resources for its own benefit, deep ecologists would question the purpose of the use of those resources and whether they were really necessary or not.

Deep ecology challenges the values of a capitalist and consumer-based society, emphasising that: 'society–nature relationships cannot be fundamentally transformed within the existing social structures' (Pepper, 1996: 21). Subsequently, it stresses the requirement for social change based upon the transformation of 'individual consciousness' (ibid.). Thus there is a need for individuals to adopt an environmental ethic that is reflected in their lifestyle and behaviour that emphasises a respect for nature. However, trying to understand what constitutes an environmental ethic is not necessarily straightforward. It implies an acceptance of a view that humans are not separate from nature but part of it. Such a view is contrary to that taken by many influential western philosophers, including Immanuel Kant and René Descartes. The separation of humans from the surrounding environment is central to Kantian philosophy which tends to be heavily influential in western decision-making (Ponting, 1991). The basis of Kantian philosophy is to treat nature as non-divine and to give humanity a spiritual freedom and domination over it. This raises three key points:

- The worth of nature is purely instrumental
- No moral limits are imposed on humanity with regard to the use of nature
- There are assumed to be no intrinsic internal limits within nature which we should respect.

Holding a similar view, the influential French philosopher René Descartes believed that animals were both insensible and irrational and could not suffer. In his view, animals were mechanical like clocks, unlike humans who possess souls and minds. Such a philosophy encourages dualism, separating humans from nature, and forms the basis for the use of nature to satisfy human desires and aspirations. Descartes encouraged the belief that humans were the masters of nature.

However, not all Western philosophers were in agreement with Descartes; for example, Benedict Spinoza believed that every being or object was a manifestation of a God-created substance. In his philosophy of 'animism' or 'organicism', there is a belief that a single and continuous force permeates all beings and things. Henry Moore, another important seventeenth-century philosopher, also believed that the spirit of god or *anima mundi* was present in every part of nature.

In contemporary times the ideas of the earlier philosophers have been refined into the discernible field of environmental ethics. Based upon the 'rights' of nature to an existence, Simmons (1993) deduces at least two possible meanings of the application of ethics to the environment. These are: (i) an ethic for the 'use of the environment', which may be summarised as adopting an anthropocentric viewpoint over how we decide to use the resources of earth; and (ii) an ethic 'of the environment', in which all non-human beings are given the same moral standing as that of the human species.

The first of these interpretations is probably the one most development decision-makers would relate to and base their judgements upon. For example, a hypothetical question could be, 'Is it ethically right for people to make their livelihoods from tourism by damaging and killing coral reefs, thereby denying the next generation the right to use this resource to earn a livelihood?' The inference of this question expresses a concern for the next human generation, rather than a concern for the living organisms of the reef, or a recognition of any intrinsic right for them to exist.

Simmons' second position is slightly more abstract, accepting that non-human beings possess a residual intrinsic goodness and moral equality with humans. Such a position would imply that the non-human world has 'equal' rights with those of the human world. The Romans, for instance, believed in *jus animalium*, the idea that animals possess rights independent of human civilisation. However, these rights did not extend to species of the wider non-human environment, with a great deal of deforestation of the Mediterranean basin taking place during Roman times.

> **THINK POINT**
>
> Should the environment be used in an 'instrumental' way to maximise economic benefits and financial profits of tourism? Can environmental management and technologies solve the environmental problems that can be created through tourism? Do we need to have a stronger environmental ethic that recognises an independent value of nature?

Although the application of ethics to tourism is advancing in the literature (e.g. Butcher, 2003; Fennell, 2006), application of environmental ethics to the use of natural resources for tourism is limited.

However, as Nash (1989) observes, certain groups of
people will benefit from the denial of ethics to nature, as illustrated
in Box 2.9.

Such a direct reference to the culling of seal cubs would probably be
regarded as unethical and a denial of animal rights by many in society.
Perhaps this represents the consumption of the environment in its most
direct form through tourism. In return for payment, tourists are buying
the right to kill living animals, just as tourism based upon game hunting
continues to be encouraged in certain parts of the world. Although
culling can be defended from an anthropocentric viewpoint by the need

Box 2.9

Tourists rush for kill-a-seal pup holidays

Governments in Canada and Norway have decided to allow tourists to participate in the killing of baby seals as part of annual culling programmes.
The justification for this action is that there are too many seals, resulting in
a decline in fish stocks. Following the government ruling, tour operators
decided to promote holidays based on sealpup culling. The Canadian decision was taken in the early 1990s, with package holidays being marketed
for US$3,000 in America, to buy the 'right' to club seal pups to death on
the Newfoundland ice in the annual culling programme. The packages were
popular as Mike Kehoe, the executive director of the Canadian Sealers
Association commented: 'People want to come out and kill and it's a good
market for us' (Evans, 1993).

In Norway, the involvement of tourists in seal culling was set to begin in
January 2005, with one company NorSafari advertising culling trips on the
internet. Jowit and Soldal (2004: 3) comment: 'The company's website
shows photos of hunters posing with their kill and offers trips that not only
include accommodation and food but help with cutting up and preserving
seal carcasses.' Although professional seal hunters have traditionally used
clubs, tourists would kill the seals by shooting, a presumably more humane
but expensive way of execution. The Norwegian Fisheries Minister said that
the move would restore the ecological balance between fish and seals along
Norway's coast, although environmental groups argue that over-fishing is
the cause of devastated fish stocks and not the seals.

to sustain fish stocks, the selling of this type of holiday based upon the aggressive instincts of tourists, raises ethical questions over human actions towards the animal world.

Conversely, the utilisation of the conservation ethic may entail a denial of human rights. For example, although the creation of protected areas to conserve nature and wildlife may seem positive, this may result in the displacement of people from the resources that they rely upon for their livelihoods, leading to accusations of 'eco-fascism'. Such criticisms of the establishment of national parks in developing countries have been made on the basis of their exclusion of local people, as Leech (2002:78) comments:

> Though many of us crave sanctuaries away from the modern world, the idea of wilderness areas where tourists can sample nature, free of man, is a western romantic illusion. And further, what is evident is that the preservation of these areas comes at a very high cost to local communities which are either removed from the land, have their lives regulated or are forced to play the role of theme park extras to satisfy the demands of ecotourism.

At the beginning of the twenty-first century it would seem that ethical concerns over the interrelationship between humans and the environment are beginning to manifest themselves in the tourism sector. The extent to which these ethical concerns will influence the processes of development decision-making is at the moment largely unknown and will only become evident with the passage of time. However, the development of ethical codes of conduct by internationally recognised organisations, such as the United Nations World Tourism Organisation (UNWTO), does suggest that ethics will have a more prominent role in tourism decision-making in the twenty-first century than they had in the last century.

Summary

- The term 'environment' can have different meanings, including 'our surroundings'; 'objective systems' of nature, e.g. coral reefs, rainforests; and our 'perceived surroundings'. How we view the environment and our relationship to it is shaped by religious and cultural belief systems.
- Understanding why people become tourists is complex but is related to lifestyle characteristics of urbanised societies in advanced stages of

capitalism. Tourism can be viewed as a form of 'conspicuous consumption', relaying cultural messages about lifestyle and identity. Tourist behaviour in destinations will not be uniform, and will be representative of a mix of a range of individual motivations with wider cultural forces, which shape beliefs and attitudes. Subsequently, how tourists experience the natural environment will not be homogeneous and a variety of behaviour patterns will be displayed.

- What are regarded as being 'desirable' and 'beautiful' landscapes for tourism reflect processes of economic, social and cultural changes in society. The combination of Industrial Revolution, urbanisation and the Romantic movement were highly influential in making coasts and mountain areas popular destinations for recreational tourism.

- The dependence of tourism upon the physical and cultural attributes of destinations raises ethical questions over who are the beneficiaries of tourism and the 'rights' of the non-human environment. This ethical questioning reflects concerns over the existing equality of relationships between different stakeholders in tourism and between the human and non-human environment.

Further reading

Butcher, J. (2003) *The Moralisation of Tourism: Sun, Sand . . . and Saving the World?*, London: Routledge.

Fennell, D. (2006) *Tourism Ethics*, London: Routledge.

Krippendorf, J. (1987) *The Holiday Makers*, Oxford: Heinemann.

Leopold, A. (1949) *A Sand Country Almanac*, Oxford: Oxford University Press.

Nash, R.F. (1989) *The Rights of Nature: A History of Environmental Ethics*, Wisconsin: The University of Wisconsin Press.

3 Tourism's relationship with the environment

- Changing perceptions of tourism's relationship with the natural environment
- The negative consequences of tourism for the environment
- How tourism can aid conservation

Introduction

As discussed in the last chapter, the rapid growth in demand for international tourism experienced in the second half of the twentieth century has resulted in the raising of ethical concerns over how the cultural and physical environments of destinations are used for tourism. This chapter considers the range of positive and negative consequences tourism can have upon the environment.

Changing perspectives on tourism's relationship with the environment

The reliance of tourism upon the natural and cultural resources of the environment means invariably that its development induces change which can either be positive or negative. There is an increased awareness of the environmental effects of tourism amongst governments, non-governmental organisations (NGOs), the private sector, donor agencies, academics and the public. This interest is reflective of a marked change in attitudes to our interaction with the environment that has occurred during the second half of the twentieth century and into the twenty-first century. Society's concerns after the Second World War rested primarily with rebuilding the economies of Europe and environmental priorities were subsequently low. By the late 1960s, as the effects of the pursuit of economic growth upon the environment became more evident, environmental issues began to gain more prominence. The first breakup of a major oil-tanker, the *Torrey Canyon* in 1967, leading to the release of oil on to the south-west coast of England, caused a high level of public concern, and highlighted the fact that increased living standards were not free of environmental risk. This was the first time anything like this had ever happened, as the oil polluted the water and washed on to the beaches. Media images of birds with black oil and tar stuck in their feathers which stopped them from flying, were poignant. So too was the knowledge that the oil had killed fish and shrimps and other forms of life. In the same year, the first major oil spill in the United States occurred from an offshore rig near Santa Barbara, releasing millions of tonnes of crude oil on to the coasts of California. In 1969, toxins leaked into the River Rhine, poisoning millions of fish and threatening the quality of drinking water for millions of Europeans (Dalton, 1993).

In the same decade as these 'environmental disasters', the increasing industrialisation of farming in the USA was also heavily criticised in Rachel Carson's (1962) book, *Silent Spring*, for the ecological damage caused by the use of agro-chemicals on farmland. This book subsequently had a major influence on public consciousness and eventually on regulatory policy, with the banning or restriction of use being placed on 12 of the pesticides and herbicides that Carson had identified as being most dangerous, including the notorious DDT.

Subsequently, there was a growing body of evidence that industrial growth and progress did not come free of environmental cost. In 1968, perceptions of the world as having unlimited and abundant resources were also challenged by the first widely broadcast television images of the earth shot from the American spacecraft Apollo 8, showing the earth as a sphere floating in space. The concept of a 'spaceship earth' was the subject of a famous essay in environmental studies by Boulding (1973), questioning the 'cowboy economy' associated with the reckless and exploitative use of nature, which he believed typified the western approach to development. In place he argued that we should begin to conceptualise the earth as having a 'spaceman economy'. From this viewpoint the earth, like a spaceship, doesn't have unlimited reserves of anything, and humans must find their place without threatening its cyclical ecological system.

However, tourism remained largely immune from environmental criticism, the image of tourism being predominantly one of an 'environmentally friendly' activity, the 'smokeless industry'. This perception was enhanced by the imagery of tourism, embracing virtues of beauty and virginity, as portrayed in landscapes of exotic beaches and mountain areas framed in sunshine. Nevertheless, there were one or two dissenting observations about the 'smokelessness' of tourism. Milne (1988) comments that in 1961 there was concern being expressed over the possible ecological imbalance that could result from tourism development in Tahiti in the Pacific. The observation of the effects of increasing numbers of tourists in Europe in the 1960s led Mishan (1969: 141) to write:

> Once serene and lovely towns such as Andorra and Biarritz are smothered with new hotels and the dust and roar of motorised traffic. The isles of Greece have become a sprinkling of lidos in the Aegean Sea. Delphi is ringed with shiny new hotels. In Italy the real estate man is responsible for the atrocities exemplified by the skyscraper approach to Rome seen across the Campagna, while the annual invasion of tourists has transformed once-famous resorts, Rapallo, Capri, Alassio and scores of others, before the last war no less enchanting, into so many vulgar Coney Islands.

By the 1970s people were becoming more aware and concerned over environmental issues. In 1972, the results of research by a group of scientists and business leaders into population growth, resource use and other environmental trends were published in the *Limits to Growth: A Report for the Club of Rome Project* (Meadows *et al.*, 1972). Predictions in the report of pollution, resource depletion and heightened death rates resulting from a lack of food and health services raised public consciousness over environmental issues. In 1979, the near meltdown of the nuclear reactor at Three Mile Island in Pennsylvania, and the subsequent threat of a major environmental catastrophe, alerted the public to the dangers of the civil nuclear power programme. Opposition to nuclear power became a central focus of the environmental movement in the 1970s, based upon both the environmental consequences of the programme and its strong link to the development of nuclear weapons. During the 1970s, questions about the environmental impacts of tourism began to be raised more widely, as tourism expanded into new geographical areas and the negative effects of its development became more obvious.

Recognition of the problems that could be caused by tourism led the Organisation of Economic Co-operation and Development (OECD) to establish in 1977 a group of experts to examine the interaction between tourism and the environment. Negative effects on the environment from tourism such as the loss of natural landscape, pollution, and the destruction of flora and fauna were already being noted. These concerns were also expressed in academic circles, with the publication of *The Golden Hordes* by Turner and Ash (1975), in which the whole process of tourism development was polemically questioned as exemplified in the following passage:

> Tourism is an invasion outwards from the highly developed metropolitan centres into the 'uncivilised' peripheries. It destroys uncomprehendingly and unintentionally, since one cannot impute malice to millions of people or even to thousands of businessmen and entrepreneurs... As a mass movement of peoples, tourism deserves to be regarded with suspicion and disquiet, if not outright dread.
>
> (Turner and Ash, 1975: 127)

Similarly, the negative impacts of tourism were lamented by Goldsmith (1974:10):

> A large part of the coast of Southern Spain, of the South of France and of the Italian Riviera have already been mutilated beyond redemption with countless hotels together with their associated amenities. An island that has suffered particularly from tourism is Hawaii. This once beautiful island is disfigured with countless skyscrapers.

By the 1980s, global environmental problems resulting from human agency had begun to become popular media items. There was a subsequent raising of a wider consciousness of environmental issues, including global warming and ozone depletion. Concern was also being increasingly and vociferously voiced over the depletion of the tropical rainforests of the world for agriculture and logging. Nuclear power remained a concern, as Europe experienced its worst nuclear disaster at Chernobyl in the Ukraine in 1986, with the effects of the nuclear fall-out being experienced across several European states.

During the 1980s, the spread of mass tourism beyond the Mediterranean basin into new areas, including South-East Asia, Africa and the Caribbean, meant that there was an increasing focus on tourism as a form of economic development in developing countries. Besides economic aspects of development, this focus also included concern over the environmental and cultural consequences of tourism development. The awareness that tourism could have negative effects was increasingly being recognised by NGOs. Pressure groups including Tourism Concern, the UK-based campaigning group for humane tourism development, and the Ecotourism Society in the USA were established in the 1980s to promote ethically based tourism for both indigenous peoples and nature. Local pressure groups, concerned by the effects of tourism development on their physical and cultural environment, were also formed, such as the Goa Foundation in India. The Goa Foundation has opposed tourism because of the loss of access to resources for local people and other associated human rights violations associated with its development. There was also evidence of increasing dissatisfaction by tourists with areas that were perceived as being overdeveloped or having lost their original attractiveness. For instance, Barke and France (1996: 302), commenting on the problems facing tourism in the Costa del Sol region of Spain in the late 1980s, state: 'Environmental decay and poor image have combined with overcrowding and low safety and hygiene standards, together with the popularity of cheaper forms of accommodation and catering, to reduce the perceived attractiveness of the region.'

In the 1990s, new environmental concerns became prominent, reflecting both local and global concerns. An ethical dimension was increasingly introduced into environmental campaigning over the rights of non-human life, with high-profile and sometimes violent actions being taken for the liberation of animals from experimentation. Protests against road building became a central focus for environmental campaigners in Britain and other European countries, as concerns over the loss of countryside and nature grew. The enthusiasm and popular support for this campaign led the British government to a major rethink of its strategy over transport, particularly the role of the private motor car and road building, leading to the cancellation of numerous road-building projects. Green politics in Europe gained increasing recognition through democratic political routes in the 1990s, notably the formation of a governing red–green coalition in Germany. Indeed, by the end of the decade green politicians were in charge of the environment ministries of Germany, France, Italy and Finland (Bowcott et al., 1999). Concerns over the practices employed by farmers were also heightened with the outbreak of bovine spongiform encephalopathy (BSE) in Britain, which not only threatened animal life, but could also be transmitted to humans in the form of Creutzfeldt-Jakob Disease (CJD). Worries over genetically modified crops were also raised in Europe and there was a subsequent increased demand for organically produced vegetables, fruits and meats. Major protests were also made over the global inequality in world trade and the role of the World Trade Organization in encouraging the removal of trade barriers and import tariffs.

By the end of the 1990s, tourism development had for the first time been attacked directly by eco-warriors. Ski facilities were burnt down in Vail in Colorado at the beginning of 1999 because of their possible impacts upon wildlife. In the 1990s, some tour operators, hotels and airlines began to take action to mitigate their negative environmental impacts, as is discussed in Chapter 7. Consideration of the role of tourism in conservation and poverty alleviation led NGOs and government agencies, including the World Wide Fund for Nature (WWF), Oxfam, the Netherlands Development Organisation (SNV) and United Kingdom's Department for International Development (DFID) to engage with tourism. A growing number of tourists similarly became more interested to varying degrees in the environmental aspects of tourism as green consumerism became more popular. Alternative types of tourism, including 'ecotourism' and 'sustainable tourism' became established in the tourism vernacular. A summary of the changing attitudes of western society towards the environment and tourism over the last five decades is shown in Box 3.1.

In the first decade of this century, the relationship between tourism and the environment is more hotly debated than ever. The term 'sustainable' has become integrated into government policy and industry's strategies. The emphasis on stakeholders' responsibilities to the natural environment has transcended beyond those of government and industry to include consumers. This is exemplified through the debate on tourism's 'carbon footprint' and the extent to which it is deemed unethical to fly more than a certain number of times per year.

Whilst there is evidence to suggest that tourism can act as a negative force of change upon nature, it may also act as an agent of conservation. The vitriolic discourse that sometimes appears to be levelled against tourism needs to be tempered with consideration that the use of natural resources brings economic and social benefits to destinations, and may aid human development. This can be critically important for regions of developing countries where alternative development options may be very limited and people struggle to meet their basic needs for food, clean water and shelter. It may also be the case that in certain situations, tourism may offer a less environmentally damaging alternative for development than other options. However, this is not to condone the negative effects of

Box 3.1

The relationship between society, environment and tourism

Decade	Attitudes to the environment	Attitudes to tourism
1950s	Instrumental use for wealth creation	International tourism still restricted to a relatively small elite; high levels of participation in domestic tourism
1960s	Heightening environmental awareness; *Torrey Canyon* oil disaster; publishing of Rachel Carson's *Silent Spring* attacking farming practices in the USA	Quickening pace of 'mass' participation in international tourism; few expressions of concern about the environmental consequences of tourism development
1970s	Growing awareness of pesticide and fertiliser pollution associated with farming; concerns over water pollution; publication of the Club of Rome report *Limits to Growth* in 1972; awareness of global pollution and global warming in scientific circles; Three Mile Island nuclear power meltdown in Pennsylvania; formation of 'Greenpeace' in Canada in 1971	Increasing awareness in academic circles that tourism is not a 'smokeless industry'; Organization for Economic Cooperation and Development establishes a working committee on tourism and the environment
1980s	Issues such as 'global warming', 'acid rain' and 'ozone depletion' begin to gain media coverage. Chernobyl nuclear power accident in the Ukraine; concern over the loss of the tropical rainforests; origins of green consumerism; Brundtland Report in 1987. Intergovernmental Panel on	Continued growth and spatial spread of tourism to South-East Asia and the Pacific; mass tourism in the Caribbean; by the end of the 1980s tourist arrivals began to fall to traditional locations such as the 'costas' of Spain, which were seen as *passé* and overdeveloped; tourism increasingly viewed as a development tool for less developed countries. Founding of tourism pressure groups such as Tourism

Continued

Box 3.1 continued

Decade	Attitudes to the environment	Attitudes to tourism
	Climate Change (IPCC) established in 1988	Concern (UK) and the Ecotourism Society (USA)
1990s	Protests against: development and road building; genetically modified crops; animal experiments; loss of rainforests; inequalities in world trade. On-going global concerns and an increased propensity to purchase organic food. First United Nations Conference on Environment and Development (Earth Summit) held in 1992. Kyoto Agreement to control global emissions agreed in 1997	'Eco-warriors' target tourism development in Colorado. More tourists becoming environmentally aware. The tourism industry begins to respond to concerns over the environment. 'Eco-tourism', 'green tourism', and 'sustainable tourism' become popular phrases
2000	Global warming becomes an issue of global concern. Media coverage is heightened; the scientific community is virtually united in its view that the global temperature rise is a consequence of human activities. International agreements are sought on carbon reduction schemes. Carbon taxation begins to be proposed by some national governments. Kyoto Agreement comes in force in 2005	The contribution of aviation to global warming receives increased press coverage. Airlines establish carbon offset websites for voluntary donations from customers. Growing acknowledgement in the tourism industry and government that climate change will threaten the success of some tourism destinations, especially small islands susceptible to a rise in sea-level, and lower-altitude downhill ski resorts in which snowfall is expected to become marginal.

Source: After Hudman (1991)

tourism upon the environment; as in many situations, these result from a mixture of human ignorance and greed, rather than from a philanthropic desire to improve the living conditions of human beings.

Our knowledge of the environmental impacts of tourism is still limited. This is a consequence of the amalgam of the following factors:

- Research into impact studies is relatively immature and a true multidisciplinary approach to investigation has yet to be developed.
- Research into the environmental consequences of tourism tends to be reactive and therefore it is not always easy to establish a baseline against which to monitor changes.
- It is not always easy to separate out the environmental impacts attributable to tourism from the effects of other economic activities or anthropogenic factors, such as human habitation, and non-anthropogenic causes, such as natural environmental change.
- It is not always possible to separate the source of impacts upon the environment between local residents and tourists.
- The consequences of tourism are difficult to assess because tourism development is often incremental and the effects are cumulative.
- Spatial discontinuities are inherent to tourism. For example, the effects of air pollution caused by air and car emissions associated with tourism may contribute to acid rain, which destroys forests hundreds of kilometres away.

(After: Hunter and Green, 1995; Mieczkowski, 1995)

The impacts of tourism upon the natural environment can be separated into two broad categories of negative and positive changes. To provide a structure to the discussion, the first part of the following section deals with the negative consequences of tourism and the second part with the positive effects. This ordering is reflective of much of the observation and commentary, which has raised awareness of the negative environmental aspects that can result from tourism development, whilst the positive environmental aspects have traditionally been less well defined.

The negative impacts

There are a broad range of negative environmental impacts resulting from tourism development. These may be categorised into three major types of concern: natural resource usage; behavioural considerations; and pollution. These are summarised in Box 3.2.

Box 3.2

The negative environmental consequences of tourism

Issue	Problems	Examples
Resource usage: tourism competes with other forms of development and human activity for natural resources, especially land and water. The use of natural resources subsequently leads to the transformation of ecological habitats and loss of flora and fauna	Some natural resources that tourism relies upon have characteristics of Common Pool Resources (CPRs). Thus there is a propensity for overuse. Indigenous and local people can be denied access to natural resources upon which they base their existence and livelihoods. Land transformation for tourism development can directly destroy ecological habitats and ecosystems. The use of resources for tourism involves an 'opportunity cost', as they are denied to other sectors of economic development	• Airport construction in tourism generating and destination areas such as London and Malta uses large areas of farmland • Draining of coastal wetlands in Kenya for hotel developments • Loss of beach and coral reef ecosystems in the Caribbean • Deforestation of mountainsides associated with tourism in the European Alps and Himalayas • Lowering of the water table below the level of local wells as in Goa, India • Induced change to ecological habitats and a subsequent reduction in the number of species of flora and fauna as in Scotland and the European Alps
Human behaviour towards the destination environment	Local people encouraged by the revenues to be gained from tourism, and tourists, may display ignorance and/or a disregard for the environment and indulge in inappropriate behaviour. This can lead to a range of consequences for the physical and cultural environments	• Disruption to eating and breeding patterns of wildlife animals in the Maasai, Kenya • Local people breaking off coral to sell to tourists off the Mombassa coast • Dynamiting of fish in the Amazon to provide entertainment for tourists • Tourists walking over coral in the Caribbean • Increased crime, prostitution and drug taking in many destinations • Offence caused in Muslim cultures by western tourists wearing inappropriate dress to visit mosques and other cultural sites

Box 3.2 continued

Issue	Problems	Examples
Pollution • Water • Noise • Air • Aesthetic pollution	A range of different types of pollution can result from tourism. These impact on different spatial scales from the local to the global. In destinations the effects of pollution are often associated with the level of tourism development and the degree of implementation of planning and environmental management controls	• Problems of human waste disposal generated by tourism in the Mediterranean and the Caribbean • Air pollution problems in the European Alps and the contribution of jet engine emissions to global warming and ozone problems • Noise pollution of air balloons in the Serengeti Park in Africa • Many coastal areas such as in parts of the Mediterranean and the Caribbean have had their coastlines transformed by standardised construction of tourist accommodation and are indistinguishable from each other

Natural resource usage and tourism

Ecosystem issues

Tourism involves different stakeholders with a variety of interests in and expectations of the natural environment. They include the tourism industry, government, tourists and local communities, all of whom will be searching for some kind of benefit from tourism, including the financial, economic, experiential and social, the quest for which may place pressure upon natural resources. Some of the natural resources that tourism relies upon possess characteristics of what in environmental studies are referred to as 'common pool resources' (CPRs). They are characterised by criteria of exclusion and exploitation, where exclusion is impractical on the basis of cost or is at least very costly, and the exploitation of the resource by one person reduces the benefit for another (Ostrom et al., 1999). A major threat to the well-being of CPRs exists from the mentality of 'finders-keepers'; that is, a rush to harvest and secure the benefits of the resource before someone else does (Hardin, 1968). Typical CPRs used for tourism include the oceans, the atmosphere, beaches, coral reefs, wetlands and mountains. For a more comprehensive discussion of CPRs, refer to Box 4.5 and a summary of Hardin's (1968) seminal essay, 'Tragedy of the Commons'.

Alongside the threats of pollution to the oceans and atmosphere that are discussed later in the chapter, a focus for tourism are coral reefs. They are the second most biologically diverse ecosystem on the earth after tropical rainforests, being home to approximately 25 per cent of all marine species despite covering only 0.17 per cent of the ocean floor (Goudie and Viles, 1997), and subsequently are a great attraction for tourists. Their growth requires specific environmental conditions, including a water temperature of between 25 and 29 degrees centigrade, a relatively shallow platform of less than 100 metres below sea-level to grow upon, highly oxygenated water, and areas free from sediment or pollution. There are three main types of coral reefs: 'fringing reefs', which connect directly with the shore; 'barrier reefs', which are separated from the shore by a lagoon; and 'atolls', which consist of a ring of coral reefs around a lagoon. The reefs are composed of a thin veneer of living corals that exist on the top of the skeletons of previous biological growths.

Coral is placed under threat from different aspects of tourism, including the construction of tourist facilities, inadequate sewage disposal measures to deal with human waste, and the behaviour of local people and tourists. Coral reefs have been mined for building materials in Sri Lanka, India, Maldives, East Africa, Tonga and Samoa (Mieczkowski, 1995). Besides being used for building materials, reefs may also be placed under threat from the dust created by construction which gets blown on to the reef, for example, in the Red Sea off the shoreline of Egypt. According to Goudie and Viles (1997), 73 per cent of the coral reefs off the coast of Egypt are thought to have been adversely affected by tourism.

The addition of untreated sewage into the water causes nutrient enrichment or eutrophication, stimulating the growth of algae, which can cover coral reefs, in effect suffocating them. For instance, the discharge of partially treated sewage into the sea off the Hawaiian island of Oahu stimulated the growth of the alga *Dictyosphaeria cavernosa*, which overgrew and killed large sections of the reef (Edington and Edington, 1986). Similarly parts of the Great Barrier Reef off the coast of Australia have been destroyed by the proliferation of 'crown-of-thorns' starfish, *Acanthaster planci*, which feed on the reef and have become abundant as a consequence of pollution and over-fishing of their predators. The future of the Great Barrier Reef is also under threat from global warming (IPCC, 2007a).

Box 3.3

Coral reefs and game parks: issues of tourism development in Kenya

The country of Kenya in East Africa is endowed with natural resources that should make it an attractive tourism destination for decades into the future. It has fantastic game reserves, a beautiful coastline with coral reefs, and cultural diversity. Nevertheless, some of these resources are being put under threat from tourism development. The most popular area for holiday tourism in Kenya is along the coast, north and south of Mombassa. The coast has a diverse range of ecosystems including coral and mangrove swamps. However, the development of tourism has placed the coral under threat from sewage pollution from hotels, tourists walking on the coral and killing it, local people breaking it off to sell as souvenirs to tourists, and boats dragging their anchors through it. Mangrove swamps which form the basis of another vital ecological chain are also being cleared in an unsustainable fashion. They are being cut for construction poles, timber and fuel wood, as well as being removed for aquaculture farms. As a result of the demand by tourists for lobsters, crabs, prawns and fish, there has been over-fishing and a decline in fish stocks, and subsequently lobsters and prawns now have to be imported from Tanzania.

Nor is tourism faring much better in some of the safari parks in Kenya, tourists being attracted by the 'Big Five' game animals: elephant, rhino, buffalo, lion and leopard. Indigenous people were displaced to create the parks in the 1940s, notably the Maasai people from the Maasai Mara park. This has radically altered their way of life and resulted in some of the Maasai having to migrate to the coast to sell their handicrafts to tourists. The consequences for the wildlife of increasing numbers of tourists in some of the parks have been the disruption of their breeding and eating patterns. Tourists are taken too close to the animals, elephants have been poisoned from eating zinc batteries thrown out in the rubbish from the tourist lodges that have been established, and there is also erosion of the vegetation from the excessive numbers of minibuses crossing the parks threatening the grazing of herbivores. Other problems in the parks include the pollution of water through lack of adequate sewage and other waste disposal facilities. This has led to wild animals and vultures feeding on the waste food and drinking from open sewers.

Sources: Visser and Njuguna (1992); Monbiot (1995); Badger *et al.* (1996)

The behaviour of local people and tourists towards the reefs can also damage the coral, as is discussed later in the chapter. One country in which tourism has resulted in negative effects upon its coral reefs, and other types of natural resources is Kenya, as is described in Box 3.3.

However, the blame for the demise of the world's coral reefs cannot solely be laid at the door of tourism, and its overall impact is not thought to be large compared to other industrial sectors. Tourism is a contributory factor to the destruction of coral reefs along with other natural and human causes. These include storms and hurricanes, El Niño, over-fishing, increased sedimentation produced by the deforestation of land, and industrial pollution. Hurricanes can kill living coral by flinging it up to the top of the reef, whilst El Niño (which is an ocean current occurring off the coast of Peru every two to ten years) causes changes in ocean currents and sea temperatures. Changing sea temperatures have also been attributed to the effects of global warming, as discussed in Chapter 8. The raising of sea temperatures can threaten the food supply of coral, which is dependent upon microscopic algae called *Zooxanthellae* for turning the sun's energy into sugars, to provide energy for the coral polyps. The algae are very temperature sensitive, and as the temperatures of the ocean rise, the algae will leave the polyps hence denying them food, which may subsequently lead to the mass death of the corals. It is thought that with continued global warming and increasing ocean temperatures, this process of coral bleaching may occur more regularly. Additional threats for coral are also posed by more aggressive kinds of human actions. The reefs of French Polynesia in the tropical Pacific have been threatened from nuclear bomb testing by the French, whilst the residual radiation remaining on some of the atolls from the American and British nuclear tests in the 1960s means they cannot be used for habitation or tourism (Mieczkowski, 1995). Industrial activity can also threaten the well-being of coral; for instance, increased silting as a consequence of dredging for tin poses a threat to coral reefs in Thailand (Jenner and Smith, 1992).

The destruction of coral reefs not only means a loss of biodiversity but also threatens other environmental resources upon which tourism depends. In the same way that tourism was likened to a spider's web in Chapter 1, major natural features such as coral form part of interlinked and complex ecosystems. Coral not only provides an abundance of food for fish, but also acts as a natural breakwater to help protect coastline and beach areas from erosion. Without this natural breakwater, beach erosion will occur at a much quicker rate than normal, especially where palm trees, the roots of which help to stabilise the beach, are removed to

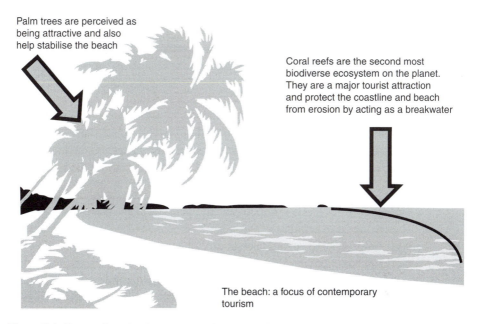

Palm trees are perceived as being attractive and also help stabilise the beach

Coral reefs are the second most biodiverse ecosystem on the planet. They are a major tourist attraction and protect the coastline and beach from erosion by acting as a breakwater

The beach: a focus of contemporary tourism

Figure 3.1 *Pre-tourism development: the ideal paradise*

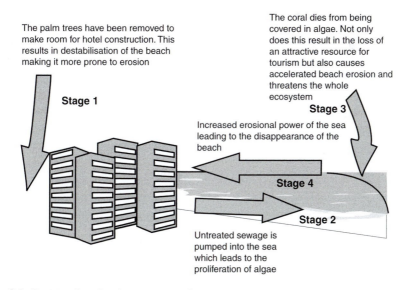

The palm trees have been removed to make room for hotel construction. This results in destabilisation of the beach making it more prone to erosion

The coral dies from being covered in algae. Not only does this result in the loss of an attractive resource for tourism but also causes accelerated beach erosion and threatens the whole ecosystem

Stage 1

Stage 3

Increased erosional power of the sea leading to the disappearance of the beach

Stage 4

Stage 2

Untreated sewage is pumped into the sea which leads to the proliferation of algae

Figure 3.2 *Post-tourism development: the disaster*

make space for hotel construction. The potential damage that unaware and unplanned tourism development can do to coral and beach environments is shown in Figures 3.1 and 3.2, developed from the work of Edington and Edington (1986).

An actual example of the type of construction mentioned in stage 1 of Figure 3.2 is shown taking place in Figure 3.3, which is a photograph of tourism development on the coastline of the Indian Ocean. The construction of tourist facilities would seem to be happening at a rapid pace, leading to the removal of palm trees, and destabilising the beach area.

Another important ecosystem in terms of making the earth a habitable place to live for humans are coastal and inland wetlands, which cover approximately 6 per cent of the world's surface. Wetlands not only house a wide range of biodiversity and act as vast carbon-storing areas for the world, but also operate as a form of local flood control measure by absorbing vast amounts of water at times of high rainfall, and discharging it to adjacent areas in a slow and measured way. However, the restricted availability of land for tourism development in coastal areas, combined with the engineering ability to reclaim wetlands, means that wetlands are increasingly being used for tourism development. For instance, in the Languedoc-Rousillon region of France, five new coastal areas were

Figure 3.3 *Beach construction alongside the Indian Ocean*

developed on reclaimed wetland sites, in response to the lack of available land for building on the French Riviera (Klemm, 1992). One of the coastal areas, 'La Grande Motte', has subsequently become the third largest French resort in the Mediterranean after Cannes and Nice. The engineering project was enormous, involving the draining, dredging and filling of coastal marshes and saltwater lagoons, leading to the complete disappearance of the wetland fauna. A similar scheme to reclaim wetlands including mangrove swamps has taken place in the Caribbean, to accommodate the Sir Donald Sangster International Airport in Jamaica.

After coastal areas, the second most popular location for tourism is in mountain areas. Just as tourism has resulted in detrimental environmental effects in coastal areas, it has also caused negative impacts in some mountain ecosystems. Mountain areas are physical environments which are sensitive to change, characterised by short growing seasons as a result of the cold temperatures, and soils which tend to be thin and nutrient deficient. The combination of these factors makes subsequent regeneration of damaged vegetation difficult.

Mountain areas have become popular destinations for activity-based tourism, and mountain systems worldwide are being used increasingly for activities such as downhill skiing and snowboarding, mountain biking, paragliding, white-water rafting and trekking. One particular form of mountain tourism that is very popular is wintersports. The economic benefit associated with sports such as downhill skiing and snowboarding has meant that their development has actively been encouraged through government policy in some countries, as a means of developing upland rural areas. Although the development of downhill skiing has aided the creation of economic prosperity in many mountain areas, it has also resulted in many negative consequences for the physical environment. The effects of downhill skiing in mountain environments are summarised in Box 3.4.

The development of mountain tourism requires the construction of hotels, apartments and associated infrastructure, placing increased pressure on land resources and animal habitats. The removal of trees to create ski runs, besides resulting in a loss of habitat for wildlife, also means that rainfall falling on the mountain slopes is not absorbed in the same quantity as before. The removal of trees causes a loss of cohesion and stabilisation of the soil by the tree roots, and subsequently the mountain slope is more prone to slippage. The combined effects of increased amounts of water running across the surface of the slope, its weakened stability, and the force of gravity, has led to mountain areas becoming

Box 3.4

The effects of downhill skiing upon the environment

Type of development	Processes	Results
Piste preparation	Removal of vegetation and boulders to a depth of 20 cm to allow snow accumulation	• Ecosystem damage, e.g. loss of Arctic-alpine vegetation • Visual pollution – loss of aesthetic quality, especially in the summer
	Deforestation of the mountainsides	• Increased avalanche risk • Increased propensity for mud slides • Disturbance of wildlife, e.g. Black Grouse in the French Alps, Ptarmigan and Red Grouse in the Scottish Highlands
Lift installation	Early resorts built roads up the mountainsides to transport pylons	Ecosystem disruption – destruction of vegetation; disturbance to wildlife and loss of habitats
	Use of heavy cables to support tows and chairs	Death of birds colliding with cables, e.g. Black Grouse in the French Alps; and the Red Grouse and Ptarmigan in the Scottish Highlands
Artificial snow making equipment	Increasing use of artificial snow cannon which involves great water usage, e.g. to produce one hectare of skiing surface requires 200,000 litres of water	• Increased water usage – diversion of water and lowering of the water table • Energy consumption • Noise pollution • Use of additives to aid crystallisation of the water into snow, leading to contamination of the soil
Increased infrastructure development	• Building of extra roads to transport skiers • Hydro-electric schemes	• Land use transformation; noise and air pollution • Increased levels of salination causing loss of flora e.g. the Australian Alps
Superstructure development necessary for destinations	Construction of hotel development and other usual amenities such as cafés, restaurants, bars	• Land use change • Air and water pollution

vulnerable to landslides. The effects of these landslides can be quite dramatic, sometimes involving the loss of human life. For example, Simons (1988) describes the avalanches of mud that swept down mountainsides during the summer of 1987 in northern Italy and southern Switzerland, in which 60 people died, 7,000 were made homeless and 50 towns, villages and holiday centres were wrecked. The cause of these landslides was attributed to the removal of mountainside forest for ski development.

In the developing world, tourism is also having an impact in mountain areas. For example, in the Annapurna area of Nepal, tourism has been developed as a means of alleviating poverty and aiding conservation. However, whilst it has helped enhance economic and social well-being, there exists controversy over the extent that it has caused environmental problems, notably deforestation. The main cause of consternation rests upon the difficulty of disaggregating the effects of tourism upon natural resources vis-à-vis those caused by local people.

Competition for natural resource usage

It is axiomatic that tourism is dependent upon the use of natural resources, consuming them both for development and as attractions. Sometimes, competition for resources may lead to conflict between different users. Thinking of tourism as a system, one of the most noticeable developmental aspects of tourism in generating and destination areas is airport construction. Airports are an essential part of the international tourism system and can generate major employment opportunities for local people. The expansion and development of airports is also beneficial to the tourist by offering easier access to a wider choice of destinations. However, the resultant effects for the environment, including the people who live near to airports, are not always as beneficial. For instance, airport development and expansion often involves the transformation of agricultural and recreational land which is covered by runways and terminal buildings. According to Friends of the Earth (1997), major international airports, like Heathrow in London, have paved areas equivalent to 320 kilometres of three-lane highways or motorways.

In destination areas the extensive amount of land used for both airport and seaport development can also be problematic. For example, in 'small island developing states' (SIDS), the loss of agricultural land for airport and seaport development can lead to an increased reliance upon food imports to meet local needs (Briguglio and Briguglio, 1996). The development of an airport also requires additional infrastructure, such as new roads and railways, which again necessitates land-use changes and adds to the pollution of surrounding areas. As the demand for international tourism increases, the demand for the expansion of airports is likely to grow. According to Whitelegg (1999), air transport demonstrates the biggest growth rate of any form of transport and the International Air Transport Authority (IATA) estimates that the rate of growth in air passengers will continue at 5 per cent per annum until the year 2010.

Within destinations, the development of the tourism superstructure, such as hotels, attractions and its associated infrastructure, also requires land. Tourism is often a competitor for land use with other economic activities, such as agriculture, and in some cases extractive industries like logging and mining. Consequently, the use of land for tourism development is at the denial of other forms of economic activity, thereby incurring what economists refer to as 'opportunity costs'; that is, the potential economic benefits resulting from another type of development other than tourism are denied. The danger of unplanned and unregulated tourism development, in response to market conditions in which there is a high demand for tourism, can mean there is an overuse of resources for tourism and a lack of development of other forms of economic activity. This can lead to an economic overdependence upon tourism and a lack of diversification of the economic base. The danger of this situation is that if tourism demand to a destination decreases, there is a lack of development of other economic sectors to support the local economy. This is likely to lead to high levels of unemployment and associated social problems.

Another key natural resource that is essential for tourism is water. The addition of hundreds or thousands of bed spaces in a destination, combined with the lifestyle demands of western tourists, such as a daily requirement for showering, clean sheets and bath towels, means that tourism is responsible in some destinations for the consumption of copious amounts of water compared to the needs of the local population. Salem (1995) remarks that 15,000 cubic metres of water will supply 100 luxury hotel guests for 55 days, whilst the same amount will supply 100 nomads or 100 rural farmers for three years, and 100 urban families for two years. The effects of the development of tourism in areas where water resources are limited can mean that local people are denied the access to the water resources they previously used, for example, to irrigate crops.

They may find that streams previously used for irrigation have been diverted further upstream to service tourism development, or that the water-table level has been lowered by over-extraction to service tourism establishments, rendering their wells useless. Where it is impossible to continue to extract enough fresh water locally, hotels can pay to have the water imported, whilst local people suffer water shortages. Unsurprisingly, access to water resources has sometimes led to conflict over tourism development between the developers and the local community, as exemplified in Box 3.5.

In Goa, in India, tourism development has also caused discord over resource issues between developers and local people. This has resulted in local people organising protest groups against tourism development and open antagonism towards tourists, as discussed in Box 3.6.

Besides possibly having restricted access to water resources, local people may also find that they are excluded from other areas that they used to use for natural resources and recreation, such as beach areas. A typical sign on beaches that have been privatised to accompany up-market hotel development, especially in developing countries, is like the one shown in the photograph in Figure 3.4, taken on the island of Langkawi in Malaysia. Besides adversely affecting the access of local people to

Box 3.5

Confrontation over water resources in Mexico

In Tepotzlan in Mexico, the place from which Zapata led the peasant army in the Mexican Revolution of the early twentieth century, the residents (who are mostly Nahua Indians) protested against plans to build a golf course, five-star hotel and 800 tourist villas. Apart from the fact that such a development will be exclusive of local people in terms of employment opportunities, it is calculated that the development will use up to 525,000 gallons of water a day, threatening shortages in the town. Using slogans of 'Zapata lives' and 'Land and Liberty', the locals took over the town hall periodically, and barricaded the streets with barbed wire and boulders in protest against the scheme.

Source: Keefe (1995)

Box 3.6

Clashes over resources in Goa, India

The development of tourism in Goa raises many issues over the interaction between local people and tourism. Goa is a state in western India bordering the Indian Ocean. It possesses the typical 'exotic' image of paradise for westerners, with over 65 kilometres of sandy beaches and coconut palms. The area began to develop its international tourism (it was already very popular for domestic tourism) potential in the 1980s with the first German charter arriving in 1987.

The development of tourism has been characterised by investment from outside the region and the building of large four- or five-star hotels. Unfortunately, this style of development has meant that tourism has excluded some local people from the resources that they need for their livelihoods. Nicholson-Lord (1993) gives examples of the Cidade de Goa Hotel, which built a 2.4-metre wall around a beach to deny local people access, and the Taj holiday village and Fort Aguada beach resort hotels, where guests are guaranteed water 24 hours a day whilst nearby villagers are denied access to the pipeline for even one to two hours a day. Many villagers face electricity and water shortages, with one five-star hotel consuming as much water as five villages, and one five-star tourist consuming twenty-eight times more electricity than a Goan. Many of the hotels were also built directly on the beach, damaging the dunes, and human sewage was piped directly into the water without being treated.

The extent of the feeling of exclusion of local people from the benefits of tourism has led to the growth of protest groups against tourism development, notably the Jagrut Goenkaranchi Fauz (Vigilant Goans' Army) and the 'Goa Foundation', an environmental group. Occasionally there has been open aggression towards tourists, such as the pelting of German tourist buses with rotten fish in the late 1980s, and ten tourists were beaten up by a group of villagers in Nuven village after knocking down two pedestrians.

---

resources, sand from beaches may be used as material for hotel construction, ultimately leading to its disappearance. For example, in Antigua in the Caribbean, miles of pristine sand have been removed by sand-mining companies for use in constructing tourism projects, and sand has also been shipped to the Virgin Isles to build other beaches (Pattullo, 1996).

PRIVATE AREA
HOTEL GUESTS ONLY
KAWASAN PERSENDIRIAN
PENGHUNI~PENGHUNI
HOTEL SAHAJA

Figure 3.4 *Privatisation of beach areas may result in local people being denied access to resources they previously enjoyed*

Tourism development can also lead to the displacement of people from their homes, particularly of poorer people in less developed countries, many of whom possess no rights of land ownership and have limited or no access to legal representation. One of the most notable examples of the displacement of indigenous people associated with tourism was the exclusion of the Maasai people from their traditional lands, when the Maasai Mara reserve in Kenya was established in the 1940s. However, the exclusion and displacement of local people from lands for tourism development is not something that is specific to Kenya. The development of the Chitwan National Game Reserve in the Terai area of Nepal was also achieved by the exclusion of local people, and on the island of Langkawi in Malaysia, the compulsory reclamation of land by the state government led to the splitting up of a long-established community and a loss of livelihood for many villagers (Bird, 1989). The case of Langkawi is discussed in detail within the context of environmental planning and management in Box 7.2 of Chapter 7. Displacement of people from the land for the development of 'golf villages' in South-East Asia has also occurred. The development of these golf villages usually involves the transformation of agricultural land, the development of condominiums,

conference centres and other leisure facilities, with their size being up
to 80 times the size of a typical European golf course (Burns and
Holden, 1995).

Human behaviour towards the environment

An integral part of the tourism system is tourists and local people. The
behaviour of both groups will be highly influential in determining the extent
to which the consequences of tourism upon the non-human world are either
negative or positive. The behaviour of tourists to the culture of the
destination they are visiting will also be influential in determining whether
tourism is viewed as a positive or negative force for change by local people.

A major natural attraction for tourists is wildlife but it can be adversely
affected by certain aspects of human behaviour. The viewing of wildlife
species in their natural habitats has become an attractive activity for an
increasing number of tourists, resulting in the intrusion of humans into
environments which had previously been the exclusive preserve of
wildlife. Ironically, the desire of tourists to enhance their understandings
of nature by observing wildlife at close quarters can bring disruption to
the natural behaviour of the wildlife they want to see. According to
Duffus and Dearden (1990) cited in Roe *et al.* (1997), the extent of the
impact of tourism on wildlife can be related to the type of tourist activity
and the level of tourism development. Mathieson and Wall (1982) also
add that the resilience of wildlife to the presence of humans will
influence the degree to which tourism proves harmful to a particular
species. For instance, the type of safari tourism practised in the Serengeti
Park on the Kenyan/Tanzanian border is representative of a highly
developed level of tourism, involving local operators taking tourists into
the park in minibuses, and animals being surrounded by 30 or 40
vehicles full of tourists taking photographs. The invasion of the territorial
space of the animals, and the associated increase in noise levels, raises
the stress levels of animals, which is disruptive to their breeding and
eating patterns. For example, cheetahs and lions are reported to decrease
their hunting activity when surrounded by more than six vehicles
(Shackley, 1996). The drivers of the minibuses are encouraged to ignore
laws governing the proximity of their vehicles to the animals by the extra
tips they receive from tourists for getting close to them.

Sometimes the threat to wildlife from tourism can be more direct,
especially in situations where the level of environmental education is low,

and an insensitive view of the environment is held by locals. For example, commenting on the backpacker operations that take tourists into the rainforest in Ecuador, Drumm (1995: 2) writes:

> Only 20% of local guides have completed secondary education, and very few are proficient in a language other than Spanish. Together with a social context which embues them with a settler frame of mind, antagonistic to the natural environment, the negative impacts of this operation are significant. Hunting wild species for food and bravado is commonplace during tours as well as the occasional dynamiting of rivers for fish. The capture and trade in wildlife species including especially monkeys and macaws is also very common.

Besides wildlife, other natural resources may be placed under threat from the action of local people. For instance, coral is damaged by local people who break it off to sell as souvenirs, as in the Bahamas and Granada, where rare black coral is made into earrings for sale to tourists. Local operators taking tourists out in boats to visit reefs sometimes drag their anchors through the coral causing localised damage, whilst tourists harm the coral by touching and standing on it. Additionally, shells are sometimes collected by local people to sell to tourists, as in areas of the Red Sea, the Caribbean and off the coast of Kenya. Key factors influencing the attitudes of local people to the surrounding environment are poverty levels, economic opportunity and the extent of the provision by government and the private sector of environmental education.

Careless behaviour by tourists can also adversely affect wildlife and ecosystems. A common problem associated with tourists is littering, which can potentially result in the death of animals eating the litter, and also lead to the attraction of predators of endemic species into areas where they would not normally go. For example, elephants have been killed by eating zinc batteries thrown on to rubbish heaps surrounding the outskirts of lodges in the Maasai Mara in Kenya, as described in Box 3.3. In the Cairngorm mountains in Scotland, foxes that are predators of the indigenous ptarmigan and grouse have been enticed into the area, leading to a decline in bird numbers (Holden, 1998). However, as Shackley (1996) points out, although tourism can be detrimental to wildlife, the habitat of much wildlife is under a more widespread threat from other phenomena, such as agricultural development, urban expansion, and extractive industries like logging and mining.

The behaviour of tourists can also cause cultural changes in the societies they visit. One of the possible consequences for cultures that are exposed to tourism, especially in developing countries where there is likely to be a marked difference between the lifestyle of the tourist and the local population, is what anthropologists term the 'demonstration' effect. According to Burns (1999: 101), the 'demonstration effect' applied within the context of tourism can be explained as:

> This effect refers to the process by which traditional societies, especially those who are particularly susceptible to outside influences such as youths, will 'voluntarily' seek to adopt certain behaviours (and accumulate material goods) on the basis that possession of them will lead to the achievement of the leisured, hedonistic lifestyle demonstrated by the tourists.

Seemingly there exists a desire for some groups living in non-western cultures to imitate western lifestyles by acquiring the same symbols, a phenomenon also referred to as 'blue-jean' culture. The imitating of western lifestyles may, however, go beyond the purely superficial, to appropriating the more materialistic and individualistic values of western society. The adoption of such values will undoubtedly cause change in societies whose previous cultural value systems are likely to have emphasised religion, co-operation, the family and the community. The copying of tourists' fashions can also lead to cultural conflicts between different groups in communities. For example, post-pubescent Muslim girls living on the Mombassa coast in Kenya who see western women dressed in bikinis, may wish to dress the same way, even though their religion dictates they should be completely covered. In societies with strong traditional values, such a situation is likely to result in conflict with the elders and men of the community.

Another criticism that is often made of tourism's effect upon local cultures is that it can lead to their commercialisation and commoditisation. This can result in established rituals and ceremonies becoming a parody of the authentic culture, to satisfy the demands of tourists, in effect turning into 'pseudo-events'. Tourists are unlikely to understand the significance and meaning of the event they are watching or participating in, and with the passage of time it is also possible that the performers may lose sight of the original cultural importance of the practice (Williams, 1998). Similarly, the same criticism can be made of local handicrafts, which instead of being handcrafted for their cultural significance, can end up being mass-produced for sale to tourists. Yet as

Burns (1999) points out, it is often difficult to separate out cultural changes resulting from tourism from those attributable to the wider process of modernisation, such as the geographical spread of global television networks.

However, the material wealth and lifestyle of the tourist can have the effect of making one's own culture feel comparatively worthless and limiting. What local people fail to understand, as no one usually informs them, is that the behaviour of the tourist is atypical of how they behave during the rest of the year. The fact that western society suffers from increasing levels of family breakdowns leading to social isolation, high crime rates, drug addiction and suicides, and that there is a developing underclass of impoverished and isolated people, is seldom pointed out to people in developing countries. This was illustrated by Marcel-Thekaekarka, who had spent ten years working with the Adivasi, the indigenous people of the Nilgiri mountains of Tamil Nadu in India, on rural development schemes before she visited Germany with six of the Adivasi. Although in material poverty themselves, the Adivasi were shocked by the lifestyle of people in Germany, even though they were materially much wealthier. The Adivasi expressed sympathy for their lifestyles which seemed urbane and isolated. As Marcel-Thekaekarka (1999: 3) writes:

> 'It's very nice to be here', Chathi, one of the six told me. 'But I couldn't live here. *It's not my place.* A man needs his family, his community, his own people around him. Just money can't give you a life. You'd shrivel up and die.'

Sometimes the problems that arise from tourism are what the majority of societies would perceive as being socially unacceptable, involving crime, prostitution and drugs. For example, the development of sex tourism in Thailand and other parts of South-East Asia has resulted in increased levels of Aids amongst the population. It has also led, as is the norm in prostitution, to the exploitation of women by men.

Pollution

Pollution of the physical environment resulting from tourism can occur on different spatial levels, including tourism-generating and destination areas, and other localities not directly connected with tourism but to which pollution is displaced. However, tourism is just one contributing factor to local and global pollution, along with a whole range of other services and industries. The pollution associated with tourism may be categorised into four main types: water, air, noise and aesthetic pollution.

Water pollution

Water pollution is a major problem in many tourist regions of the world. For instance, in the most visited tourist area of the world, the Mediterranean, only 30 per cent of over 700 towns and cities on the coastline treat sewage before discharging it into the sea (Jenner and Smith, 1992). In the Caribbean Basin, where 100 million tourists annually join the 170 million inhabitants, only 10 per cent of the sewage is treated before being discharged into the sea. The most worrying aspect is that, compared to other areas of the world, these figures are good. Other regular international tourist destinations such as east Asia and Africa and the islands of the South Pacific, with a few exceptions, have either no sewage treatment or treatment plants that are totally inadequate for the size of the population (ibid.). Although nearly 20 years after the writing of Jenner and Smith, it could be expected that these figures would have improved, the extent of any improvement is likely to be irregular, dependent upon political will and economic resources to mitigate the causes of pollution. The problem of water contamination from human sewage is not caused exclusively by tourism but is reflective of an inadequate infrastructure to meet the needs of both local people and tourists.

Besides the consequences it can have for human health, causing diseases ranging from mild stomach upsets to typhoid through the intake of water contaminated by faeces, human sewage also causes eutrophication (nutrient enrichment) of the water. This can pose a particular threat to coral and associated ecosystems, as explained earlier in the chapter. Eutrophication of water may also lead to a downturn in tourism demand, as experienced on the Romagna coast of Italy in 1989. According to Becheri (1991), the total number of tourist bookings on the Romagna coast fell by 25 per cent in 1989 compared to 1988, owing to the eutrophication of the Adriatic and the spread of algae on the surface of the water. The source of the pollution was from agricultural, urban and industrial wastes, which were deposited in the River Po and subsequently outflowed into the Adriatic. Similarly, the fear of a typhoid outbreak in Salou in Spain in 1988 resulting from contaminated water led to a

70 per cent decline in tourist bookings the next year (Kirkby, 1996). By the late 1980s, many Spanish beaches were considered dirty, with only three beaches on the Costa del Sol being considered clean enough for Blue Flag status in 1989 (Mieczkowski, 1995). The decline in water and beach quality contributed significantly to a slump in tourism receipts in Spain in the late 1980s.

Besides the pollution resulting from the inefficient disposal of human waste, water pollution is also caused by fertilisers and herbicides, which are widely used on golf courses and hotel gardens. The water containing the chemicals seeps through to the groundwater lying 5 to 50 metres below the earth's surface and through aquifers it eventually reaches rivers, lakes and seas (ibid.). Other sources of water pollution are caused by motorised leisure activities such as power boating, and even suntan oil being washed off tourists when swimming can result in localised pollution. However, although tourism may seem to be a culprit of much of the planet's water pollution, it is important to realise it is a contributory factor. The major sources of water pollution come from oil spills, industrial waste pumped into the sea, and from chemicals used in agriculture.

Air pollution

A major source of air pollution within the context of tourism is associated with transport. Both air and car transport contribute to local and global atmospheric pollution through the burning of fossil fuels. The release of carbon dioxide (CO_2) is widely thought to be a major cause of global warming, and the emission of sulphur dioxide (SO_2) contributes to problems of acid rain which destroys forests and historic monuments such as the Parthenon in Athens. Per passenger kilometre, aviation produces more CO_2 than any other form of transport, as is shown in Box 3.7. These results are based upon a compilation of various reports investigating CO_2 emissions for different modes of transport (Dubois and Ceron, 2006). The figure for car transport is given per vehicle as the emission level per passenger will be determined by the number of people travelling in the car.

The growth in air travel since the 1950s has been rapid, with the annual rate of passenger growth averaging 5 to 6 per cent per annum, over almost a 50-year period. Expressed in actual numbers, in the period between May 2006 and May 2007, there were an extra 114,000 flights and 17.7 million extra passengers (McCarthy, 2007). It is this rapid

Box 3.7

Emission factors for comparative modes of transport

Type of transport	Emission factor (kg CO₂ – equivalent per passenger km)
Plane: Mid haul	0.432
Plane: Long haul	0.378
Train	0.026
Bus	0.019
Car	0.18 (per vehicle km)

Source: Dubois and Ceron (2006)

growth in air transport and its increasing significance in global CO_2 emissions that is of concern. This growth is fuelled by the cheap flights sector but also by rapid increases in certain countries. In China domestic flights increased by 18 per cent and international flights to and from the country by 17 per cent between May 2006 and May 2007 (ibid.). Besides contributing to global warming, air transport emits nitrogen oxides, which are believed to reduce ozone concentrations in the stratosphere (Friends of the Earth, 1997). Emissions of nitrogen oxides and hydrocarbons at lower levels also contribute to regional smog problems by forming low-level ozone on calm summer days, which is harmful to health.

An easy misperception is to equate transport in tourism solely with airlines, as it is often assumed that most tourism is undertaken by using air travel. However, this is not the case; for example, within Europe, the car accounts for 83 per cent of the total passenger kilometres (Cooper *et al.*, 1998). A very common pattern of summer holiday travel in Europe is for tourists from the countries of northern Europe such as Germany, Scandinavia and the Benelux countries to drive down to the Mediterranean coast for their vacation. When domestic tourism is also taken into account, then the effect of the motor car becomes even more prominent, as this form of transport constitutes the majority of domestic trips. For those people living in transport transit areas, the effects of tourism are predominantly ones of inconvenience, associated with pollution and safety concerns. Although the social and health effects of transit traffic upon local communities is an under-researched area,

Zimmermann (1995: 36), commenting on transit traffic through the European Alps, remarks: 'The transit traffic is one of the most evident problems within the Alpine area. In several regions local populations' endurance levels have already been reached or exceeded.'

Within destination areas the air quality may deteriorate as a result of both extra traffic and construction. Dust generated during the construction of tourist facilities contributes to air pollution; for example, Briguglio and Briguglio (1996) remark that the demolishing of existing buildings and the construction of new ones for tourism has generated vast amounts of dust in Malta. Yet, as with water pollution, tourism can also be adversely affected by displaced air pollution originating from sources elsewhere. Forests, used for tourism, that are situated close to large industrial centres, are particularly under threat from acid rain, caused by emissions of sulphur dioxide from coal-burning power stations. For instance, Mieczkowski (1995) refers to the Black Forest in Bavaria becoming the 'Yellow Forest', and Jenner and Smith (1992) comment that as a result of the damage to the forest caused by acid precipitation, there has been a loss of tourism income. It should be noted that just as tourism counts as a contributory factor with a range of other human activities to water pollution, the same is true for air pollution. The use of the motor car in everyday life for commuting and shopping, and the burning of fossil fuels to produce electricity, are larger air-polluting activities than tourism.

Noise pollution

In psychological studies of humans, noise pollution has been found to affect behaviour in a detrimental fashion. According to Mieczkowski (1995), most complaints associated with tourism relating to noise are from air traffic. Noise pollution is particularly a problem for those residents who live around busy international and domestic airports, and the proposed development of airports may sometimes lead to violent opposition by local people and protest groups, as was the case with the construction of Narita Airport in Tokyo (Shaw, 1993).

Noise pollution from tourism will be particularly noticeable in destinations where tourists are searching for quietness and peace. Air flights in remote areas where quiet is expected, such as the Grand Canyon in the USA and the Himalayas, can cause disruption to tourists and recreationists. Noise pollution from the construction of tourism facilities can also be a problem for residents and tourists. Briguglio and Briguglio (1996) observe that intense noise is generated by the building of hotels and other construction activity in destinations. Night clubs open until the early morning, and increased car traffic from tourism movements, all add to the noise pollution experienced by both residents and tourists in tourism destinations.

Aesthetic pollution

The development of tourism facilities can also lead to a decline in the aesthetic quality of the environment. Commenting on the development of tourism in the Guadeloupe and Martinique islands situated in the Lesser Antilles, Burac (1996: 71) comments:

> The most worrying problem now prevalent in the islands relates to the anarchic urbanisation of the coasts.... Also, the built-up areas by the seaside are often not aesthetically attractive due to the diversity of architectural styles, the disappearance of traditional creole homes and the disorderly way in which public posters are displayed.

> **THINK POINT**
>
> Have you visited tourism destinations that you have found visually unappealing? If so, why did you find them unattractive?

Often tourism development is based upon maximising profits whilst ignoring aesthetic concerns. This has led to a uniform style of development along many coastlines of the world that ignores local architectural styles, building traditions and materials as is shown in Figure 3.5.

> **THINK POINT**
>
> Harm to fauna and flora resulting from tourism development is an emotive topic. Yet it can bring economic benefits to people who live in poverty, such as the money to buy medical supplies and food. To what extent is a reduction in biodiversity important (especially if aesthetic appreciation is not harmed) if economic benefits are being brought to an area from tourism?

In mountain areas, tourism has also created unsightly development. Besides hotel and apartment construction, the development of ski lifts and pistes has also been heavily criticised as a form of aesthetic pollution. For instance, the Scottish Office (1996: 7) makes the following remarks about the development of downhill ski facilities in Scotland: 'In addition [to adverse ecological effect], the infrastructure and uplift facilities associated with skiing can have a visual impact on what would otherwise be an unspoilt and undeveloped landscape.'

The positive effects

It is unlikely that any kind of human action has a beneficial effect for the natural environment it interacts with, other than to protect or conserve it from more damaging forms of human behaviour. Therefore, when we talk about the beneficial effects of tourism for the environment, we are in essence talking about tourism being used as a way of protecting the environment from possibly more damaging forms of development activity, like logging and mining. The exception to this situation is tourism's interaction with the built environment, where in post-industrial landscapes, tourism has been used as a catalyst to aid urban regeneration and improve the quality of the environment.

Figure 3.5 *Tourism development may result in a uniformity of structure that fails to reflect local building styles, as here in Lanzarote*

The development of tourism will, however, normally place an increased emphasis on the maintenance of a 'good-quality' environment in a destination, if tourism is intended to play a long-term role in the local economy. Yet the meaning of what is a 'good-quality' environment is highly value-ridden and debatable. Nevertheless, it is certain that the long-term economic success of tourism is often dependent upon maintaining a level of quality in the natural environment, which will satisfy the demands of tourists. As Mieczkowski (1995: 114) comments: 'The very existence of tourism is unthinkable without a healthy and pleasant environment, with well preserved landscapes and harmony between people and nature.' The consequences for tourism destinations that do not maintain a high-quality environment were illustrated by the examples of Salou in Spain, and on the Romagna coast in Italy, cited earlier in this chapter. This relationship between the economic success of tourism, the environment and the tourist is shown in Figure 3.6.

This diagram emphasises that the environment, including both its cultural and physical resources, is the key to satisfying the needs of the tourists and building sustainable economic prosperity for tourism. It is therefore in the long-term interest of the destination community to ensure that the landscape remains well preserved and that they provide stewardship of the environment.

Importantly, tourism can play a role in the conservation of the environment by giving it an 'economic value' through the revenues from tourist visitation, as is discussed in Chapter 4. Given that development decisions are predominantly based upon an economic rationale that has failed to reflect the full economic costs of development, the revenues from tourism can help to protect habitats and wildlife from other, more environmentally harmful forms of development, such as mining and logging, or from other forms of destructive human activity, such as poaching. The economic rationale of conservation through

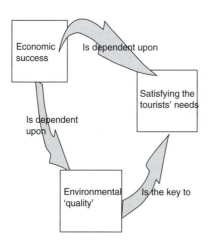

Figure 3.6 *The relationship between the natural environment, the local economy and tourism*

Box 3.8

Gorillas in Rwanda

One of the most cited examples of how tourism can be used to aid conservation is in Rwanda in central Africa. The Parc National des Volcans in Rwanda is home to more than 300 of the world's estimated 650 mountain gorillas. They were particularly under threat from poachers, an export trade of gorillas' hands for ash-trays to the Middle East was one lucrative outlet, and also from the encroachment of agriculture leading to the removal of their habitat.

The gorillas were popularised by the film *Gorillas in the Mist*, about the life of Dian Fossey, who was an active researcher and conserver of the gorillas, ultimately resulting in her murder. There is little doubt that the film aided the development of tourism in the park. Tourist visitation of the gorillas is controlled by the Office Rwandaise du Tourisme et de Parcs Nationaux, and part of the revenues generated from tourism goes to conservation agencies, notably the Mountain Gorilla Project and Dian Fossey Gorilla Fund. Visitors are taken out in small groups (a maximum of eight) by well-informed local guides to see the gorillas in their natural habitat of dense bamboo forest. The total annual visitation to the park is between 5,000 and 8,000 tourists. Importantly, the project has been proven to be a financial success, critically benefiting local people, who in turn have taken a more active interest in conservation.

However, the civil war in Rwanda between 1990 to 1994 unfortunately brought visitation to a standstill. In 1994 when 750,000 refugees were escaping from Rwanda to Zaire (now named the Democratic Republic of Congo), tens of thousands of people per day passed through the park, bringing with them their cattle and belongings. Many took refuge in the park. The occupation posed particular problems for the gorillas, including land-mines being placed in the forest, and gorillas being trapped in response to a lucrative black market for meat that had developed in response to famine. Between 1994 to 1998 rebel militias still held out in the park. The future of gorilla tourism in Rwanda therefore remains uncertain because of political unrest.

Sources: Mieczkowski, (1995); Shackley, (1995, 1996); Lanjouw, A. (1999)

tourism is a theme explained in detail in the next chapter. Within the context of this chapter, one of the most positive examples of where tourism has been used to conserve a particular species as described in Box 3.8 is the mountain gorilla project in Rwanda, which was proving very successful until the outbreak of civil war that took place between 1990 and 1994.

One type of environment, in which some of the best examples are to be found of tourism making a positive contribution to environmental improvement, is in post-industrial urban environments. Although tourism can bring problems of traffic management, overcrowding and associated problems, such as littering in urban areas, it can also play an active part in the restoration of redundant industrial areas and of historic sights. Many of the post-industrial towns of the western world suffered the loss of traditional manufacturing industries, such as iron and steel, shipping and coal-mining in the 1980s. Subsequently, tourism was turned to by many local governments as a catalyst for urban regeneration and as a means of providing new employment opportunities.

The pioneer of using tourism to regenerate urban areas was Baltimore in the USA, where the decision was made in the 1980s to rejuvenate the waterfront through the development of shopping and recreation, to arrest inner-city decline. Despite the numerous commendations that have been made of the redevelopment of Baltimore through tourism, Law (1993) sees the success as being only superficial, suggesting that poverty and housing problems remain in the city. However, there is little doubt that environmental improvements have resulted from the development of tourism in depressed urban areas. This is particularly the case for the waterfront areas of major cities, such as Sydney in Australia, and Liverpool in England.

One advantage of improving the environmental quality of the urban environment, especially where it is combined with improved infrastructure development, is that it enhances the image of the city and makes it more probable that other businesses and services will be attracted to relocate and invest there. Besides the advantage of attracting secondary investment, tourists who come to the area also spend money, which induces a multiplier effect, generating further demand for goods and services in the local economy.

The redevelopment of urban areas through tourism can also be aided by the development of tourist attractions that are rooted in the local heritage and history of the area. For instance, at Wigan in England, the town that was the source of inspiration for George Orwell's novel *The Road to Wigan Pier*, the local municipality developed a heritage centre called 'Wigan Pier'. Reflecting the day-to-day life of an industrialised Wigan at the beginning of the twentieth century, the centre attracts over 500,000 visitors per annum. Importantly, the involvement of local people in establishing the centre helped regenerate local pride and interest in their heritage (Stevens, 1987).

The development of heritage attractions to attract tourists to urban areas has now become a global phenomenon; for example, Holden (1991) remarks that in Singapore a large historical and cultural theme park based upon the ancient Tang Dynasty of China has been developed, including 1,000 replicas of the Terracotta Warriors. However, the development of such attractions has led to criticisms of inauthenticity, especially where the theme seems to have little to do with the area in which it is situated, or the attraction attempts to recreate life as it was in a historical period.

As with all forms of tourism, the benefits of urban tourism will only continue to be enjoyed if it is carefully planned and managed to ensure that areas do not become totally over-congested, as in the case of Venice, which is literally sinking under the weight of tourists. The large influx of tourists has been a contributory factor to deterioration in the quality of life for Venetian residents, which has led to some of them leaving the city (Page, 1995). As for all forms of tourism, urban tourism has threshold limits beyond which the environment will be perceived as having declined in quality, with consequences for both local residents and tourists.

Summary

- The awareness of the environmental consequences of tourism has grown as society has become more environmentally conscious. Once perceived as being the 'smokeless industry', the expansion of tourism globally has led to an increased questioning of its environmental effects. Of particular concern is the expansion of air transport and its increasing contribution to global CO_2 emissions and subsequent global warming.
- Tourism can have negative impacts upon the environment. Major issues of concern rest over resource usage, pollution, and aspects of tourist behaviour towards the environment they are visiting.

The negative effects upon the environment include both physical and cultural aspects. However, these negative effects must be offset against the economic benefits offered through tourism. These may be of significant importance in combating poverty and aiding human development in developing countries.

● Tourism can help protect the environment from potentially more damaging forms of development, such as logging and mining. It can have a particularly beneficial role in the regeneration of economically depressed urban environments.

Further reading

Hunter, C. and Green, H. (1995) *Tourism and the Environment: A Sustainable Relationship?*, London: Routledge.

Mathieson, A. and Wall, G. (1982) *Tourism: Economic, Physical and Social Impacts*, Harlow: Longman.

Mieczkowski, Z. (1995) *Environmental Issues of Tourism and Recreation*, Lanham, MD: University Press of America.

Suggested websites

International Ecotourism Society: www.ecotourism.org

Tourism Concern: www.tourismconcern.org.uk

United Nations Environment Program: www.unep.org

World Wide Fund for Nature: www.panda.org

Tourism, the environment and economics

- The relationship between economics and the natural environment
- Issues of economic growth, common pool resources and externalities
- How tourism can be used to conserve the environment using an economic rationale

Introduction

The fact that tourism can have negative impacts upon the environment, including the use of natural resources in an unsustainable fashion, the creation of pollution and the displacement of peoples, suggests that the wealth creation from tourism development is not universally beneficial. The processes of wealth creation and resource allocation are of particular interest to economists, and concern over the effects of development upon the environment has led environmentalists and some economists to criticise free-market economics as being an inefficient mechanism of resource allocation. Tourism as a means of wealth creation is reliant upon the use of natural resources, and subsequently this chapter examines how the relationship between the environment and tourism can be understood from an economic perspective.

The role of the natural environment

Most environmentalists and economists recognise that the natural environment provides certain key services for society, as outlined in Box 4.1.

Box 4.1

Services provided by the natural environment for society

- Resources for wealth creation, based upon renewable and non-renewable types
- Ability to act as a waste disposal system, assimilating the wastes of industrial production and other activities such as leisure and tourism
- Influences our well-being by providing us with aesthetic appreciation and landscapes for recreation
- Provides us with a life support system, that is, the combination of different ecosystems provide us with oxygen and water without which we could not exist.

Sources: After Turner *et al.* (1994); Willis (1997)

Within the context of tourism, all of these services have a direct relevance. The natural attributes of destinations provide the resources for wealth creation, by attracting tourists to them. Tourists desire to experience a 'good-quality' environment through participation in tourism as was discussed in Chapter 2, thus endowing them with a sense of well-being, through aesthetic appreciation and the provision of relaxing surroundings. The environment also performs the function of a waste disposal system for the tourism industry; for instance, sewage is discharged from hotels into the sea, and aircraft release engine emissions into the atmosphere. At the same time the physical environment provides the communities of tourism destinations, and tourists, with the oxygen, water and food resources that are necessary for their survival.

It was the threat that economic development based upon free-market principles was posing to these services that led to criticisms of the ability of conventional economics to reflect the full environmental costs of production and consumption (Mishan, 1969; Turner *et al.*, 1994; Booth, 1998). Traditional models of economics are based upon continued growth, achieved by raising levels of production and consumption, on the assumption that this process will bring increased benefits to society. The process of consumption provides individuals with a sense of well-being or what economists refer to as 'utility'. However, this system of production and consumption fails to take into account the rate of depletion of environmental resources, and ignores the effects of pollution on resources that are not under private ownership, and therefore cannot be directly costed into the market system, such as the oceans and the atmosphere. Consequently, there is an inability to account for the long-term effects of consumption upon resource depletion, including the denial of the use of resources to future generations for their own wealth creation. This closed-system approach, now challenged by many economists themselves, can therefore be criticised as failing to acknowledge the mutual dependency between the human and non-human environment.

A further criticism of the conventional economic model is that it ignores basic scientific principles, relating to the first and second laws of thermodynamics (Turner *et al.*, 1994; Booth, 1998). The first law of thermodynamics states that energy and matter can neither be created nor destroyed. The second law states that when resources are processed, the amount of useless energy (entropy) increases. A common form of entropy is pollution, and how to account for the costs of pollution upon the environment is a key question currently facing the discipline of economics. The major criticisms levelled against conventional economics are highlighted in Box 4.2.

Economic growth and human welfare

The phrase 'economic growth' has become a familiar one in contemporary society, and one that would seem to be politically

---

Box 4.2

Criticisms of the conventional economics approach to the environment

- Separates society's pattern of production and consumption from the natural environment and therefore ignores levels of resource usage
- Fails to account for the long-term effects of the depletion of natural resources
- Ignores the thermodynamic law of entropy and therefore fails to integrate the full costs of pollution into the market mechanism.

desirable. However, the effects of the processes used to achieve economic growth on the environment have led to increased questioning of how it should be achieved. The paradigm of 'economic growth' can be traced to the Cambridge economist Arthur Pigou, and the publication of his thesis, 'Economics of Welfare' in 1920. In his thesis, feelings such as happiness and satisfaction were referred to as 'welfare', and welfare could, according to Pigou, be expressed and measured in cash terms. Any satisfactions that could not be measured in cash terms via the market mechanism were to be ignored (Douthwaite, 1992). Pigou's analysis was therefore confined to 'economic welfare', alternatively expressed as happiness and satisfaction, which could be measured in monetary terms. Subsequently, one method of achieving happiness and satisfaction that is measurable in cash terms is through the consumption of goods and services. However, because the feelings of fulfilment and satisfaction we may gain from enjoying the environment do not necessarily have a measurable cash value, as a monetary exchange is often not involved to acquire them, they have often been marginalised by decision-makers.

Critically, Pigou also equated the amount of economic welfare with the national income per head, thereby establishing a link between increasing levels of measured wealth and increasing levels of happiness and satisfaction – a paradigm that still holds sway in many societies nearly 100 years later. Consequently, one of society's primary concerns is increasing the 'standard of living' and maximising individual utility. The conventional view of the way to achieve this objective is through higher levels of production and consumption, commonly referred to as 'economic growth', on the premise that this will bring higher levels of welfare and happiness.

Given that many countries in the world are faced to varying degrees with a range of economic and social problems, including high unemployment, lack of foreign exchange, poor health and education facilities, and growing populations, economic growth has come to be identified as a way of improving people's welfare. Many of the countries with the most extreme problems are usually categorised as being 'developing countries', based upon an economic comparison with the more developed countries of the west. The term 'west' is used here in a political rather than geographical context to indicate countries that meet the defined economic criteria of the United Nations to be considered as developed countries.

The economic situation of many developing countries was worsened by the dramatic fall in prices during the mid-1980s of traditional exports such as copper and tea, and the necessity to repay very high levels of foreign debt to the World Bank and other private banks. The World Bank has since acted to partially ease the debt burden of developing countries. Much of this debt was associated with borrowing money for large-scale civil-engineering projects such as dam building for electricity provision, which were expected to bring 'economic growth' via the 'trickle-down' effect, that is, money being passed from the higher to the lower echelons of society. Large development schemes were also undoubtedly linked to political goals associated with national prestige. Loans were made by governments and private banks to developing countries on the basis of the rising commodity prices that had been experienced until the mid-1980s. However, the effects of political corruption among powerful elites in many countries, the failure of many schemes, combined with the inability of a global market system to provide a more equitable distribution of income and wealth, means that several decades after the United Nations First Development Decade of the 1960s, world poverty remains a major issue.

Within the context of economic growth, tourism may have a significant role to play in wealth creation. As was discussed in Box 1.1, tourism had been used by General Franco in Spain in the 1960s as a means of generating economic growth. This view was reinforced by the expenditure of American soldiers in Thailand for 'recreational' purposes during the Vietnam War in the 1960s and early 1970s. Tourism's ability to bring economic benefits and enhance political stability makes it an attractive development option to governments of countries whose environmental assets are conducive to the demands of western tourism markets, and where other options for economic development are limited.

Following the example of Spain, a strong positive correlation has seemingly been established between the development of tourism and economic growth. Subsequently, from the perspective of government, the primary concern with tourism has been as a vehicle for economic growth. The types of economic benefits that tourism can bring to destination areas are well charted in tourism literature (e.g. Mathieson and Wall, 1982; Bull, 1991; Sinclair and Stabler, 1997; Cooper *et al.*, 1998).The most important ones from a national perspective are foreign exchange earnings, which are essential for the buying of necessities such as foodstuffs and medical supplies; reduction of the trade deficit; employment creation; increased expenditure and monetary flow; the strengthening of linkages to other sectors of the economy such as agriculture, fishing and construction; and to aid diversification of the economy from an overreliance on primary commodities.

Owing to the fact that tourism can be developed with comparatively little financial investment in comparison to other kinds of industry, it is an attractive type of development for developing countries. The rationale of a government policy for tourism has therefore traditionally been based upon the economic benefits, with relatively little consideration of the effects of tourism upon the environment. This is reflected in the economic analysis of tourism, as Sinclair and Stabler (1997: 160) comment:

> The economics literature has concentrated on estimating income and employment generation and foreign currency earnings which developing countries gain from international tourism and the implications for their balance of payments.... The studies have not considered the impact of environmental deterioration on demand and thus income, employment and currency receipts, or estimated the social costs of tourism's role in economic development, either environmental or other.

The questioning of the costs that tourism development can have upon nature is part of a wider question facing society about what it views as the acceptable costs of economic growth. This questioning has never been as vehement as in the early part of the twenty-first century, although by the early 1990s, Douthwaite (1992) had drawn attention to the confused thinking over the terminology used to conceptualise human welfare. Specifically, he commented that the terms 'standard of living' and 'quality of life' are frequently used interchangeably in contemporary society, whereas in fact they mean different things. The standard of living purely measures 'economic welfare', that is, satisfactions that can be measured in purely monetary terms, and therefore ignores a whole range of other factors that help determine the quality of life. Such factors would include the quality of the environment people enjoy, levels of cultural activity, levels of health, and the chance to develop a rewarding religious or spiritual life. Doubts over the purpose of economic growth and the pursuit of an increase in the 'standard of living', which may bear little correlation to the 'quality of life', have led some economists to question how economic success can best be evaluated.

The degree of success in achieving economic growth is traditionally measured by key quantitative indicators, notably gross domestic product (GDP) and gross national product (GNP). GDP is the value of the production of all the goods and services in a country's economy over a certain time period, usually one year (Heilbroner and Thurow, 1998). The GNP is a further refinement of the measure of GDP, including income earned by domestic residents from foreign investments, whilst deducting income earned by foreign investors in the country's domestic market during a measured period of time, typically three months or one year. Until fairly recently, GNP was the commonly used measure of a country's total production, but almost all countries now use GDP.

Both GDP and GNP are taken by governments to be key indicators of the success of economic policy, measures of the standard of living, and continue to enjoy a somewhat revered status in assuring a nation's progress. Yet both GNP and GDP tell us little about the well-being of the environment that is being used to achieve economic growth, and subsequently have received substantial criticism from environmentalists. Indeed, some economists have accused GNP of being a misleading indicator of welfare. For example, Pearce et al. (1989: 23) comment: 'But GNP is constructed in a way that tends to divorce it from one of its underlying purposes: to indicate, broadly at least, the standard of living of the population.' Similarly, Douthwaite (1992: 10) remarks: 'Since GNP only measures things which are bought and sold for cash, it ignores clean air, pure water, silence and natural beauty, self-respect and the value of relationships between people – all of which are central to the quality of life.'

A major shortcoming of GNP and GDP measures is that pollution or other types of environmental harm can contribute to their increase, for example, expenditure on services and goods required to clean up pollution and associated health care costs add to GNP. The costs of increased public spending to deal with crime in society also contribute to a growth in GNP. The term has subsequently been called *gross* national product by Porritt (1984). He adds that manufactured goods are deliberately not built to last, so that they have to be bought again, that is, they have a 'built-in obsolescence' that ignores the true costs of production to the environment.

Another criticism made of the GNP calculation is that it does not include the depreciation or cost associated with the loss of natural resources used for development. Subsequently, it is possible to increase the level of GNP whilst simultaneously destroying the environment. For example, a country that is primarily dependent upon exporting hardwood for economic growth could be cutting down trees in an unsustainable way whilst achieving a growth in GNP. The GNP figure fails to reveal the 'capital loss', that is, that the supply of hardwood may disappear in the next year or in five years' time because all the trees have been chopped down. Similarly, whilst tourism development may contribute to an annual growth in GDP or GNP, it may simultaneously be destroying the natural environment and resource base upon which it depends. As for the 'hardwood' analogy, the GDP figure fails to incorporate the 'capital loss', in this case the natural attractions and resources upon which the tourism system and industry depends.

Subsequently, attempts have been and are being made to change systems of national accounting to incorporate environmental costs and benefits. Such ideas include the System of Integrated Environmental and Economic Accounting (SEEA) and the Environmentally adjusted net Domestic Product (EDP), which attempt to include the costs of resource depletion and pollution in their methodologies. Nevertheless, as Bartelmus (1994: 35) comments: 'However, so far, there is no international consensus on how to incorporate comprehensively environmental costs and benefits in national accounts.'

Tourism and growth

The equation of growth with economic success also extends to tourism. For instance, the United Nations World Tourism Organisation (UNWTO) ranks countries in order, based upon the numbers of international tourists they receive and the total of their international tourism receipts, as shown in Box 4.3. A ranking system usually infers that those at the top are the most successful, whilst those at the bottom are the least successful.

Many ministers, particularly in developing countries, view tourism as a means of pursuing economic growth and diversifying the economy away from an overdependence upon primary products, such as cash-crop agriculture and mineral production. The success of a tourism policy has traditionally been evaluated by maintaining a growth in the number of tourist arrivals and associated expenditure, and obtaining an increasing percentage share of the total world tourism market.

In reality, this measurement of the total numbers of tourists and levels of expenditure can be misleading about the net economic benefits that tourism actually brings to a country or region. A much more accurate measure of tourism's worth to society is the amount of tourist

Box 4.3

The world's top tourism earners in 2005

Rank	Country	International tourism receipts (US$ million)
1	United States	81,700
2	Spain	47,900
3	France	42,300
4	Italy	35,400
5	United Kingdom	30,700

Source: UNWTO (2006a)

Note: Success in tourism is often 'measured' by the growth in tourist receipts, visitor numbers and international market share. However, such measurements fail to take into account factors such as economic leakages and the costs to the physical and cultural environments of the development of tourism.

expenditure retained within the local economy, the level and quality of employment generation, and the equity of distribution of economic benefits. From an environmentalist perspective, the costs to natural resources and negative externalities of tourism growth also need to be costed into economic models.

When considering the economic benefits of tourism, it is important to recognise that only a proportion of the money spent by tourists will stay in the local destination economy. Alongside the extra demand generated from direct expenditure by tourists in a destination, further income is generated in the economy from the cyclical flow of money, an effect known as the multiplier effect. Theoretically, the initial tourism investment could circulate indefinitely in the economy, but it doesn't. The reason for this is that in each round of expenditure, money leaks out of the economy, through what are termed 'economic leakages', thereby removing it from circulation. The leakage of expenditure occurs for a variety of reasons, including: repaying interest charges on foreign loans for tourism development; imports for the tourism industry, including materials and equipment such as kitchen equipment for hotels and consumables, e.g. food, drink and beach products; the employment of foreign workers who may send money home to their relatives; the repatriation of profits by foreign travel and tourism companies; the saving of money; and the paying of taxes to the government.

The degree of influence of these factors upon tourism income will largely be determined by a nation's or community's level of economic development. For example, many developing countries are reliant upon multinational corporations and large foreign businesses for the capital to invest in the construction of tourism infrastructure and facilities. Consequently, a high leakage arises when overseas investors who finance the resorts and hotels repatriate the profits to their own countries (UNEP, 2004). Larger economies also possess a more diversified structure, reducing the requirement for imports to meet the needs of the tourist industry. An additional influence upon an economy's propensity for leakages is the type of tourism market it is servicing. For example, package tours and all-inclusive holidays generally have a high leakage factor, with approximately 80 per cent of all travellers' expenditures going to airlines, hotels and other international companies, rather than local businesses or workers (ibid.). Local businesses will also see their chance to earn income from tourism reduced by 'all inclusive' vacation packages, where the tourist remains for the entire stay in their resort or cruise ship, also known as 'enclave tourism'.

When tourists pay remittances to tour operators and airlines that are foreign to the country they are visiting, economic leakage may not be solely limited to the destination, with the occurrence of what Smith and Jenner (1992) refer to as the 'pre-leakage' factor. A key determinant of this pre-leakage factor will be the use or non-use of the national airline carrier. The cost of air transport is usually a substantial percentage of the total cost of a travel vacation package, typically in the region of 40 to 50 per cent. Subsequently, if the money for this service is paid to an airline foreign to the destination, the pre-leakage factor will be high. Typical pre-leakage factors for different regions of the world are shown in Box 4.4.

From the figures given in Box 4.4, it is evident that the majority of the tourist's expenditure never arrives in the destination. For any government hoping to maximise the economic benefits of tourism, reducing the causes of leakage would be a central aim. In an attempt to reduce the pre-leakage factor, encouragement of the use of the national carrier by the tour industry and tourists is important. However, this may be especially problematic for smaller developing countries and islands, where the

Box 4.4

Proportion of a tour operator's prices that are typically received by the destination

Region	Percentage
South America	45–50
Egypt	35–50
North India	35
China	30–35
South India	20

Source: Smith and Jenner (1992)

economics of the airline business and the control of routes by established airlines dictate against the establishment of a national carrier.

Consideration of the 'real' economic benefits of tourism is important for planning the role it will have in development. It is also important for guiding judgements about the extent natural resources should be traded off for economic benefits. Thought also needs to be given by government on how to reduce economic leakages. One approach is to encourage types of tourism that are integrated with community-based services and products. Typically, this kind of scenario is equated with the development of alternative types of tourism, such as ecotourism and nature-based tourism.

Problems of externalities

The development of tourism can lead to negative effects upon the environment, which can also be expressed as costs to society. For example, the pollution of the sea caused by sewage discharges associated with hotel development may lead to the contamination of fish stocks that subsequently stop fishermen earning their living. In a similar vein, extra air and car traffic at airports could lead to increased incidences of asthma in the surrounding population. Such costs may also be expressed in an inter-generational fashion; that is, the depletion of non-renewable resources resulting from the perusal of economic benefits now, deny future generations the opportunities to meet their needs, a theme of sustainable tourism discussed in Chapter 6. With particular reference to tourism, Mishan (1969: 140) comments:

> For the cost [of the holiday] to the marginal tourist takes no account of the additional congestion cost he imposes on all others (tourists and inhabitants or of the additional loss of quiet and fresh air, or of the scenic destruction suffered by all in consequence of additional building required).

Such costs are therefore externalised by the decision-maker and are subsequently termed 'externalities'. Willis (1997: 63) defines externalities as:

> consequences (benefits or costs) of actions (consumption, production of exchange) that are not borne by the decision maker, and hence do not influence his or her actions. Negative externalities that are by-products of people's use of the natural environment are commonly termed pollution.

According to Willis (1997), when externalities exist, market demand and supply curves no longer reflect all the benefits and costs of production. There exists, therefore, a discrepancy between the 'private costs' of production and consumption, and the full 'social costs' upon society. In this case the market mechanism fails to behave in an efficient fashion. Externalities arise because many environmental 'goods', such as clean air, water and landscapes, can be classified as 'public goods' sharing characteristics of collective consumption and non-exclusion. Collective consumption means that the consumption of the good by one person does not diminish the amount consumed by another person. Non-exclusion means that one person could not exclude another from consuming the resource, hence inferring there is an open access to it, and no price can be charged for it. For example, there is little economic incentive for airlines to contribute to the protection of the ozone layer and reduce their profit margins if it continues to be treated as a resource with zero cost. As Mieczkowski (1995: 160) comments in specific relation to tourism:

> They [developers] treat the environment as an inexhaustible gift of nature because our free market system has been, so far, unable to incorporate the cost of natural resources and the value of environmental damage into the prices of tourism products. In other words environmental costs have not been internalised yet.

Since Mieczkowski's observation approximately fifteen years ago, the idea of internalising negative externalities into the market system has become part of a mainstream debate about the future environmental well-being of the planet, e.g. Stern (2006); IPCC (2007). However, the realisation that external costs of production need to be internalised and reflected in the price of goods with the aim of reducing pollution has a longer history. Significantly, the Organisation for Economic Co-operation and Development (OECD) adopted the 'polluter pays' principle as early as 1972 (D. Pearce, 1993). The basic idea of the 'polluter pays' principle (PPP) is explained by Pearce (ibid.: 41) as follows: 'The PPP requires that those emitting damaging wastes to the environment should bear the costs of avoiding the damage or of containing the damage to within acceptable limits according to national environmental standards.'

However, whilst its application at a local level (for example, for the protection of rivers or streams) has proven to be possible, its application at a global level is much more problematic. Not least because of issues such as ownership, responsibility of governance and policy, and the politics and practicalities of exclusion. The implication of the 'polluter pays' approach for tourism businesses is that the costs of environmental damage would have to be internalised into their operating costs, which in turn would probably be passed on to the consumer, through augmented prices for the good or service, as shown in Figure 4.1.

To help explain Figure 4.1, an example is used of a hotel that pumps untreated sewage into the sea. In the existing market situation, position 1 represents the optimum level of usage of the sea by the hotel for sewage disposal, as determined by free-market forces. However, this situation results in pollution of the sea, causing other tourists and residents to suffer ill health, and lost

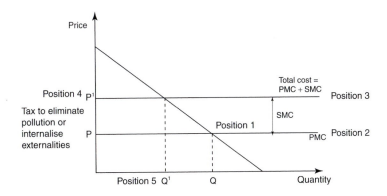

Figure 4.1 The 'polluter pays' principle

economic opportunities for fishermen as a consequence of the contamination of fish stocks. In this situation a discrepancy exists between the private marginal cost (PMC) incurred by the hotel and their guests, shown in position 2, and the total costs of production shown in position 3. The difference between positions 2 and 3 is the social marginal cost (SMC), which in this scenario represents the costs borne by third parties from the effects of pollution, in this case the fishermen and other tourists and residents.

It is the gap that exists between positions 2 and 3 that the 'polluter pays' principle suggests should be closed. One way to rectify this situation is for the government to impose an environmental tax on the hotel, or alternatively if a government has determined environmental standards for waste disposal that have been breached by the hotel, a punitive fine. This tax or fine could be estimated on the costs incurred for compensating fishermen for the lost commercial opportunities and for treating the illnesses of tourists and residents resulting from swimming in the sea. The imposition of a fine on the hotel means that the marginal external costs of the hotel's operations are now internalised to the hotel. The producer, in this case the hotel, is in effect now paying for the compensation of fishermen for their loss of opportunity to fish, and for the health care necessary to treat residents and other tourists. The hotel management may decide to pass these additional costs on to their hotel guests by raising their room rates, shown in Figure 4.1 as an increase from P to P^1, in position 4. In a price-sensitive and highly elastic market that typifies recreational tourism, the quantity of demand would shift from position Q to Q^1 (position 5), in effect meaning fewer tourists staying at the hotel and therefore less waste being pumped into the sea. This does not mean that untreated sewage does not continue to be pumped into the sea from the remaining guests, but that the level of pollution caused is now one that is socially desirable.

Similarly, Shaw (1993) uses the example of 'noise insulation' around airports to illustrate how the 'polluter pays' principle can be employed in connection with air transport. The costs of insulating houses against noise pollution from aircraft, for example, with double glazing, may in some cases be recovered from airport authorities. These costs would be borne by the airlines and probably their customers, thereby making the polluter pay. However, as Shaw points out, such an arrangement does not compensate for the loss of enjoyment of gardens or open spaces caused neither by pollution nor for the annoyance factor when a window is opened in a house, although it can be argued that cheaper house prices around airports provide some compensation. The 'polluter pays' principle is also the theme of the debate concerning whether airlines should be made to pay environmental taxes to combat global warming, which would ultimately be passed on to their passengers. It is likely that at some point an environmental tax on aviation fuel will be the first universally applied levy in the tourism industry.

One of the problems faced with the 'polluter pays' principle is valuing the environmental resources that are being damaged and subsequently being able to estimate the cost of pollution. One way around this is for governments to establish national environmental standards, and the cost of meeting these standards should in the first instance be borne by the emitter of the waste. The polluter would then most likely pay through investment in waste management systems, or in some cases, for example, if a hotel pumps an excessive amount of pollution into the sea as was the case in Figure 4.1, by a one-off punitive fine. Another method of operating the 'polluter pays' principle is through taxation, commonly referred to as 'green taxation'. According to Cairncross (1991), Pigou proposed the idea of taxation as a method of closing the gap between private and social costs, which he saw as the cause of environmental damage, back in the 1920s.

Another problem of the 'polluter pays' principle lies in establishing the 'cause and effect' relationship. Shaw (1993) remarks that it may be difficult to isolate the cause of an environmental problem, for example, atmospheric pollution may be the result of emissions from many different sources. Distance factors are also a problem, as the source of the pollution may be hundreds or thousands of miles away from where the pollution manifests itself. A further problem highlighted by Shaw is that some adverse effects are difficult to reassess in an objective fashion. However, to date environmental taxation has been noticeably absent in its application to tourism. A key element against environmental taxation from an industry perspective is the high price elasticity that is especially associated with recreational tourism. Thus, the argument is that besides harming overall demand, unless an environmental tax were applied universally, it would inevitably lead to unfair market competition, for example, some airlines, hotels, tour operators or destinations would be forced to factor these costs into their prices whilst others would not.

Public or collective consumption goods

The fact that public goods are non-exclusive and therefore have zero cost means that a market for such goods cannot operate. Subsequently, they are at threat from overuse and degradation. In tourism, many of the negative impacts of its development are related to resources to which there is open access, and where there are no, or few, enforceable legal controls on usage. The theme of overuse of public goods is discussed in Hardin's (1968) seminal essay 'Tragedy of the Commons', summarised in Box 4.5, which was important for encouraging debate about human interaction with the environment.

The emphasis in Hardin's essay is placed on human selfishness, which leads to an overuse of the resources for the pursuit of private gain. An alternative way of expressing this concept is that the 'wants' of individuals go beyond their 'needs'. According to Max-Neef (1992) cited in Pepper (1996), fundamental needs are similar in all cultures and include subsistence, protection, affection, understanding, participation, creation, leisure, identity and freedom. Conversely, 'wants' are created by the advertising industry and are attached to the acquisition of material goods and services that provide enjoyment and a means of social differentiation, in other words consumerism. Green economists would argue that the natural resources of the earth can meet these 'needs' of the human population in a sustainable way, based upon ethical principles, but are unable to satisfy all the 'wants' of the human population.

The parallel between tourism development and the scenario portrayed by Hardin is a strong one. Probably no other type of development activity is as incremental as tourism in terms of its usage of resources for development. An additional hotel here, an additional flight there, all provide extra benefits for the suppliers and consumers of tourism through increasing profits and consumer choice. However, extra supply and choice places increased pressure upon the commons, e.g. the growth of air travel and its effects upon the atmosphere.

Some critics of Hardin's (1968) work question his assumption of a finite carrying capacity. They place their faith in new technology and environmental design to extend its capacity. Others dispute Hardin's view of public attitudes to the use of the commons, arguing that common users were never oblivious to the common good (Pepper, 1996). For example, in England, herdsmen used to consult each other over the possible expansion of herds to ensure that no threat was posed to the sustainability of the commons, because it was in their interest to do so. Ultimately, finding a solution to the overuse of the earth's resources is highly complex. Not least because it relies upon an acknowledgement that those environmental problems exist, that any attempts to mitigate them will not

Box 4.5

Hardin's 'Tragedy of the Commons' (1968)

The likelihood of the overuse of common resources was highlighted in Hardin's (1968) seminal paper 'Tragedy of the Commons'. This is one of the most cited writings in environmental studies, questioning the assumption of classic economics that behaviour driven by self-interest automatically acts for the greater social good. Using the analogy of an area of common land termed 'the commons' on which farmers in a village are at liberty to freely graze their cattle, Hardin suggests that an existing state of equilibrium between the numbers of cows grazing on it and its ability to regenerate itself, can be threatened by the self-interest of the farmers. Specifically, one farmer may decide he wants to increase his herd's milk production and profits by the addition of an extra cow to his herd. Whilst this one extra cow may not directly threaten the commons' long-term stability, the other farmers witness this action and decide that they too would like to increase the size of their herds and their subsequent profits.

The farmer may also reason that the cost of the use of the resources for the extra cow will be externalised, and spread amongst all the other farmers. In this situation there are now costs being experienced by other farmers, both short term, for example, a slight reduction in the quality of the milk because the cows have difficulty getting access to enough grazing, and long term because overgrazing could lead to the loss of the commons as a resource altogether. If all the farmers subsequently decide to adopt the same position, introducing one extra cow to the commons with the aim of maximising their profits whilst externalising their costs to other producers, the commons would become ultimately overgrazed. This would threaten not only the economic viability of the existing farmers but also the ability of the next generation to use the commons for milk production.

In this situation, the true costs of production of the milk are reflected in neither the production costs nor consumer costs, because of a failure to incorporate the longer-term costs of the loss of the resource. The market failure to reflect the total costs of production means that both the producer and consumer are benefiting in the short term, from not having to pay for the long-term environmental costs of production. The overuse of the commons could subsequently lead to the loss of the resource altogether, to the detriment of the farmers and society.

Source: After Hardin (1968)

harm a nation's own interests, and a necessity for international co-operation for resource management and protection.

Tourism, economics and conservation

Criticism of the failure of conventional economics to recognise the impacts of growth upon the environment has led some economists to examine how economics can be more inclusive of the natural environment. Subsequently, attempts have been made to develop methodologies to give a value to the environment. One method is based upon expressed consumer preference for the environment, although to many environmentalists the idea of determining the environment's value in such a way may seem abhorrent, failing to recognise its intrinsic value. As Cairncross (1991: 43) comments:

> Nothing so annoys environmentalists about economists as their attempt to put a price on nature's bounty. 'What am I bid for one ozone layer in poor condition?' they tease. 'How much is the spotted owl worth?'

Putting an economic value on environmental assets is difficult but, as Cairncross continues to point out, society puts values on environmental assets all the time by deciding which policies to pursue. Given that many of the decisions about how to use natural resources are made from anthropocentric perspectives, i.e. placing an instrumental value on the environment, it is important to demonstrate the economic value of environmental assets in their existing form as accurately as possible. As Cairncross observes: 'In a world where money talks, the environment needs value to give it a voice'(ibid.). David Pearce (1993: 15) comments:

> If, on the other hand, conservation and the sustainable use of resources can be shown to be of economic value, then the dialogue of developer and conservationist may be viewed differently, not as one of necessary opposites, but of potential complements.

The willingness of individuals to give money to, and become members of, non-governmental organisations aiming to protect the environment, such as the World Wide Fund for Nature (WWF) or the World Society for the Protection of Animals (WSPA), suggests that the worth of the environment can be expressed in economic terms (ibid.). According to Turner et al. (1994), one method of aggregating individual preferences, or measuring the gains and losses of well-being is by looking at what people are 'willing to pay' for something, the most common measuring-rod of preferences being money. Turner *et al.* (ibid.: 94) comment: 'A measure of an individual's preference for a good in the market-place is revealed by their willingness to pay (WTP) for that good.' The level of willingness to pay can be assessed through direct questioning: 'What would you be willing to pay to ensure the survival of a lion as a species?' or 'What would you be willing to pay to arrest global warming?'

This technique, based upon asking the beneficiaries what they would be willing to pay for the environmental benefit under consideration, is called the contingent valuation method. As Pearce *et al.* (1989) point out, the temptation would be to respond that such species or environmental assets are 'priceless'; however, it is unlikely that many of us would be willing to give all our worldly possessions to preserve the lion. So in this sense there is an economic value attached to its preservation. One study that tried to estimate the economic value of the lion was conducted in the Amboseli National Park in Kenya, which found that each lion was worth US$27,000 per annum in 1980s values, expressed in terms of visitor pulling power. The study also demonstrated that wildlife tourism was economically preferential to the other main development option of agriculture. The park's net earnings from tourism were found to be US$40 per hectare per year, 50 times higher than the most optimistic projection for agricultural use (Boo, 1990).

Another method of calculating the value of the environment is an indirect method, the 'travel-cost method', which as the name suggests is calculated on what people spend on travelling to recreational sites such as national parks. It is indirect because it uses surrogate pricing, in this case travel costs, as a measure of value. The travel-cost method (TCM) was developed by Clawson in the United States in 1959 and perhaps represents the most widely accepted and used method for the valuation of recreational sites (Coker and Richards, 1992). The underlying presumption of the TCM is that the costs incurred in visiting, for example, a national park reflect to an extent the recreational value of the park to an individual. Information on the expenditure to a site is usually gathered as part of a visitor survey. The TCM is essentially measuring the expectation of enjoyment of visitation before the visit takes place, as the visitor or tourist has to commit to the cost of the trip to get there. In the case of tourism, given that the desires of tourists discussed in Chapter 2 related closely to vacationing in a high-quality environment, it would seem that tourists place a 'value' on the quality of the destination environment by spending money and time to travel there.

Although the revenues that can be obtained through tourism can help protect the natural environment from other more destructive development alternatives, an important caveat of this line of argument is that if natural habitats or wildlife are judged not to have sufficient economic value in comparison to other development options, then a pretext is set for their removal. Shackley (1996) also highlights another danger of the economic valuation process with particular regard to wildlife. There are many species of wildlife which are not attractive to tourists in the dramatic sense that elephants and lions may be, but have a role to play in the ecological system of the area. By their lack of inclusion, or the lack of ability to value them because they are not on the tourists' itinerary of animals to view, their value will be undermined in economic terms, placing their continued survival under threat. Shackley recommends the following approaches to valuing wildlife: calculating total gate or licence fees to estimate the value of tourism in a particular destination; estimating visitor expenditure on equipment, lodging, food and transport within a designated area; and looking at employment generated (ibid.: 127).

THINK POINT

How can tourism be used to aid environmental conservation? What are the dangers of giving wildlife a market value as a means of arguing the case for their conservation?

When tourism takes place in spatially defined areas, such as national parks or other types of protected areas that possess an evident management structure, then the use of such methodology will be easier to implement than in areas which are less well defined. Where tourists wish to see 'wildlife', then it is possible to give the wildlife a market value, based upon the willingness of individuals to pay to see it. This is exemplified in the comparative study of three national parks in different countries shown in Box 4.6.

The use of the revenues that can be obtained from tourism in national parks is varied. These would include the conservation of natural resources; funding for infrastructure; employment opportunities; supplementary livelihood opportunities; and human development opportunities. However, the vulnerability of tourism demand to external forces seriously undermines the rationale of basing protected area policy

Box 4.6

Calculating the 'willingness to pay' of visitors to national parks

The Department for International Development (DFID), in the United Kingdom, commissioned a study to examine the amount of possible revenue that was being lost from three national parks in three different countries (the Komodo National Park, Indonesia; the Keoladeo National Park, India; and the Gonarezhou Park, Zimbabwe). In all three parks the cost of entrance fees to the parks was not established by the market and it was suspected that visitors would be willing to pay considerably more to visit them. The extent of this user surplus was assessed using the 'contingent valuation' method, which explored the response of visitors to hypothetical rises in entrance fees. For each park, questions concerning tourists' 'willingness to pay' were included in tourist surveys. The following results were found:

	Proportion of sample willing to pay (%)		
Proposed entrance fee	*GONAREZHOU*	*KEOLADEO*	*KOMODO*
Current	100	100	100
X 2	79	91	93
X 4	24	70	81
X 8	8	Absent	62

Interviewees were surveyed in the park and asked how much they would be prepared to pay for the current experience. The results revealed that price elasticity (defined as the ratio of fractional or percentage change in demand to the fractional or percentage change in price) is appreciably more elastic at Gonarezhou than at Keoladeo or Komodo. However, as the authors of the report point out, the entrance fee was already five to seven times higher at Gonarezhou than at Keoladeo or Komodo. Critically the authors also found that there is a significant difference in the willingness to pay amongst different types of tourists. For example, in the analysis of Komodo National Park, members of conservation groups and older people were likely to be willing to pay higher entrance fees than any other kinds of tourists. However, the authors continue to point out the restrictions of the willingness to pay method, i.e. people are being asked how their behaviour would change in a hypothetical situation, and there is no guarantee that in reality they would behave like that.

The case study indicates that there is a potential to raise income through raising fees, and that any tourism strategy aimed at maximising revenue needs to take into account the types of visitors it wishes to attract, not interpreting the tourist market as being homogeneous.

Source: Department for International Development (1997)

upon revenues derived from it. External factors that will influence demand for tourism to protected areas include: security concerns; competition from other national parks and destination areas; and changes in market taste and fashion (Font *et al.*, 2005).

Besides methodology aimed at calculating the value of the environment, policy instruments can also be adopted to help give recognition to the value of conserving environments. One way of aiding the conservation of natural resources in developing countries is through 'debt-for-nature swaps'. The basis of this approach is that a conservation-based, non-governmental organisation (NGO) buys off some of a country's national debt, in return for guarantees that the indebted country will look after the conservation of a designated area such as a national park. This will usually involve the development of appropriate management plans for the park. In effect the NGO is purchasing the 'development rights' or alternatively the 'no-use' rights to a particular area of a country. As David Pearce (1993) points out, debt-for-nature swaps are the only way in which estimates of the 'existence' (i.e. non-use) value of nature have been established.

According to Stone (1993), the concept of the 'debt-for-nature' swap is an extension of the proposal made by Jomo Kenyatta, the founder of modern Kenya at a conference in 1961, when he suggested that if African wildlife was a world possession, then 'the world could pay for it'. The process is based upon the buying of national debts on the world's money market at a fraction of their face value; for example, an NGO may pay to a bank 25 cents for every dollar of government debt. Banks are willing to discount the value of the loan because of the risk that the indebted country may default upon its payment completely. Debt-for-nature swaps usually involve the co-operation of the national government, international and local NGOs, and banks. Essentially a government is giving up its sovereignty on a designated area of its country to an NGO. International NGOs, such as Conservation International, World Wide Fund for Nature, Nature Conservancy, and government agencies such as the US Agency for International Development (USAID), have encouraged developing countries towards conservation efforts by devoting increased attention and funding to the preservation of habitats and conservation of flora and fauna (Brown, 1998). Debt-for-nature swaps have been used in Madagascar, where the Agency for International Development (AID) paid US$1 million to purchase a part of the government's debt in exchange for the support of local environmental groups; and in the Philippines and Zambia in conjunction with the World Wide Fund for Nature (Stone, 1993). An example of the workings of a debt-for-nature swap is shown in Box 4.7.

Box 4.7

Debt-for-nature swaps: Ghana

The Central Region of Ghana possesses a rich cultural and wildlife base. Besides its indigenous culture, it was the first area of contact between European governments and the people of West Africa. Consequently, the remains of castles and forts built by the colonialists, dating back to the seventeenth and eighteenth centuries, form an integral part of the cultural tourism potential of the area. The area also possesses diverse flora and fauna, including some of the tallest trees in the world, a population of forest elephants and other endangered species including Bongos and Diana monkeys, and a 5-kilometre stretch of undeveloped beach with rolling hills and lagoons. The area possesses one of the few national parks in West Africa, the Kakum National Park.

To aid the conservation of these rich natural and cultural resources, the United Nations Development Program (UNDP) donated US$3.4 million, which was matched by the Ghanaian government. However, there still existed a shortfall of money to carry out the planned conservation of the area. Consequently, two international NGOs, the Smithsonian Institute (SI) and Conservation International (CI), in co-operation with the Ghanaian government, used the debt-for-nature scheme to raise the extra money. In effect this meant the NGOs paying US$250,000 to redeem a debt with a face value of US$1 million.

To pay for the continuing conservation of the area, and provide employment opportunities for local people, which will help to stop the poaching of the fauna for economic reasons, the scheme involves the development of tourism that is compatible with the fauna and flora of the area. Initiatives that have already begun include an audit of the resources of Kakum National Park by CI, and technical assistance for the restoration of two of the most significant castles and one fort in the locality. Interpretive trails have been developed and canopy viewing platforms constructed in the forests.

The plan also emphasises the need for community participation in shaping tourism, and a plan for the development of tourism on the beach has already been produced by local people, which includes extensive agricultural areas. Not only will these areas help to provide employment opportunities for local people but they will also strengthen links between the agricultural and tourism sectors. Local people have also been given training as tour guides and camping areas are being developed in a controlled fashion to be run by local people.

Source: Brown (1998)

Summary

- The occurrence of environmental problems resulting from economic growth has led to a questioning of the efficiency of the free-market system to account for the total social costs of production and consumption. Of particular concern is the overuse of common resources, both in the sense of exhausting supplies and the exceeding of their capability to assimilate the waste associated with production, leading to damage to ecosystems.

- Measures of society's progress are also being questioned. The use of economic indicators, such as Gross National Product, emphasises that progress can be statistically measured and that economic growth is a good thing. However, many of the things that add to our 'quality of life', such as clean air, beautiful views, family and spirituality, cannot be measured. Indeed economic growth may be harmful to these things. Success in tourism is also statistically measured by governments in terms of the numbers of international arrivals, total tourist expenditure, and world market share. However, such figures reveal little about the full environmental and social costs of tourism development.

- Being able to give an economic value to the environment can aid its conservation. Economic evaluation of the environment based upon its use for tourism can help stave off more environmentally threatening development options. The revenues from tourism can be used both to encourage the economic development of communities based in natural areas, and also provide revenues for the management of the area, in a sustainable fashion. However, the vulnerability of tourism to external factors, e.g. from economic recession, climate change and changing fashion, potentially makes conservation programmes and management financed from tourism revenues an unstable option.

- Although advances are being made in economic methodologies to cost environmental goods into the marketplace, many environmentalists would argue that nature has an intrinsic value, which is independent of humans' willingness to pay for it. Ethical arguments will remain about how humans use the environment, which cannot necessarily be resolved by costing environmental goods into the market mechanism.

Further reading

Cairncross, F. (1991) *Costing the Earth*, London: The Economist Books.

Douthwaite, R. (1992) *The Growth Illusion*, Bideford, Devon: Green Books.

Font, X., Cochrane, J. and Tapper, R. (2005) *Pay Per Nature View: Understanding Tourism Revenues for Effective Management Plans*, Netherlands: World Wide Fund for Nature.

Hardin, G. (1968) 'The Tragedy of the Commons', *Science*, 162: 1243–8.

Pearce, D. (1993) *Economic Values and the Natural World*, London: Earthscan Publications.

Sinclair, T.M. and Stabler, M. (1997) *The Economics of Tourism*, London: Routledge.

Suggested websites

United Nations Environment Program: www.unep.org

The World Bank: www.worldbank.org

5 Environment, poverty and tourism

- Understanding poverty
- Poverty's link to environmental degradation
- The relevance of tourism to alleviating poverty in developing countries
- Consideration of the limitations of tourism as a means for alleviating poverty

Introduction

The consideration by international donor agencies, governments and NGOs of how tourism may be used to reduce poverty is a relatively new addition to the use of tourism as a tool for development. The relevance of poverty in the general context of tourism's relationship with the natural environment is that it is a major cause of environmental degradation, as summarised in the Brundtland Report (WCED, 1987: 28):

> Environmental stress has often been seen as the result of the growing demand on scarce resources and the pollution generated by the rising living standards of the relatively affluent. But poverty itself pollutes the environment, creating environmental stress in a different way. Those who are poor and hungry will often destroy their immediate environment in order to survive: They will cut down forests; their livestock will overgraze grasslands; they will overuse marginal land; and in growing numbers they will crowd into congested cities. The cumulative effect of these changes is so far-reaching as to make poverty itself a major global scourge.

Therefore, poverty may be viewed as a condition that denies people the freedom to be able to take a long-term perspective on resource usage. The majority of the world's poor, defined as existing on an income per

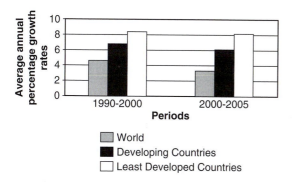

Figure 5.1 *Tourism arrivals: comparative growth rates*

person per day of less than US$1 live in the developing countries. In many of these countries, tourism has a significant economic role, for example, in 41 of the 50 poorest countries in the world, tourism constitutes over 5 per cent of GDP and over 10 per cent of exports (UNWTO, 2004). Tourism arrivals are also growing faster in developing countries, particularly in the 50 least developed countries (LDCs) than in developed countries, as is shown in Figure 5.1.

Economic development opportunities for many developing countries are limited, constrained by factors of limited natural and human resources; lack of investment flows; and the difficulty of entering into global competition with the established manufacturing and service industries of developed countries. Yet many of the developing countries, and particularly regions of countries, have a 'comparative advantage' in wilderness and wildlife that has the potential to be attractive to tourists. Importantly, these resources exist in rural areas in which 75 per cent of the world's people live in extreme poverty (ibid.).

Whilst tourism as a policy for economic development is well established, the focus upon it as a means of reducing poverty has historically received little attention. This changed in the late 1990s, as tourism's role in poverty reduction strategies received more serious consideration, notably from the United Kingdom's Department for International Development (DFID); the Netherlands Development Organisation (SNV) and (since the Millennium) the UNWTO.

This comparatively recent initiative asking how tourism can be used to reduce poverty can be linked to the wider debate about the injustices and dangers that accompany poverty. Alongside suffering and environmental degradation, suspected links between poverty and terrorism is a major concern for global security. The desire to eradicate poverty is explicit in the Millennium Development Goals (MDGs), identified by the United Nations as being critical for the future of human development. The MDGs were an outcome of the United Nations Millennium Declaration adopted by 189 nations in 2000 and are central to tackling global poverty.

The overall aim of the MDGs can be summarised as: 'a world with less poverty, hunger and disease, greater survival prospects for mothers and their infants, better educated children, equal opportunities for women and a better environment' (UNDP, 2006). The specific goals and targets are summarised in Box 5.1.

The potential importance of tourism in reducing poverty and helping to achieve the MDGs was emphasised in New York in 2005 at a summit, bringing together the UNWTO, UNICEF, UNDP and a broad range of private sector and NGOs to review the progress being made towards meeting the MDGs. These international agencies and organisations called for tourism to take its place in national development plans to help achieve the goals (Ashley and Mitchell, 2005). This recognition of tourism as a means of economic development and mitigating poverty was partially based upon the positive examples of the Maldives, Mauritius and Botswana, who have developed from Least Developed Country status on the platform of a strong tourism sector (ibid.).

However, in shaping policy and establishing goals and targets for tourism's role in alleviating poverty, the meaning of poverty needs to be given a context. Similar to the terms of 'tourism' and 'environment', the meaning of poverty defies a simple or single definition and has a variety of interpretations.

What is poverty?

Most people would identify those who are visibly starving and unable to meet their basic nutritional requirements as being in poverty. However, there would likely be disagreement over whether a person who wished to own or have access to an automobile like the rest of his neighbours, and was subsequently marginalised from the benefits that its use may directly or indirectly bring, could be labelled as being in poverty.

Inherent to this supposition are two types of poverty, 'absolute' and 'relative'. Absolute poverty refers to the inability to satisfy some stated minimum requirement which is deemed necessary to meet one's own basic physiological needs (Lister, 2004). At its most acute, absolute poverty would take the form of 'extreme poverty', meaning that households cannot meet their basic needs for survival. The sorts of conditions that characterise poor people in extreme poverty are described by Sachs (2005: 20): 'They are chronically hungry, unable to access

Box 5.1

Millennium Development Goals

Goal	Target
1) Eradicate extreme poverty and hunger	Halve, between 1990 and 2015, the proportion of people whose income is less than US$ 1 per day
2) Achieve universal primary education	Ensure that, by 2015, children everywhere, boys and girls alike, will be able to complete a full course of primary schooling
3) Promote gender equality and empower women	Eliminate gender disparity in primary and secondary education, preferably by 2005, and in all levels of education no later than 2015
4) Reduce child mortality	Reduce by two-thirds, between 1990 and 2015, the under-five mortality rate
5) Improve maternal health	Reduce by three-quarters, between 1990 and 2015, the maternal mortality rate
6) Combat HIV/AIDS, malaria and other diseases	Have halted by 2015 and begun to reverse the spread of HIV/Aids
7) Ensure environmental sustainability	Integrate the principles of sustainable development into country policies and programmes and reverse the loss of environmental resources Halve by 2015, the proportion of people without sustainable access to safe drinking water and basic sanitation By 2020, to have achieved a significant improvement in the lives of at least 100 million slum-dwellers
8) Develop a global partnership for development	Address the special needs of the least developed countries, landlocked countries and small island developing states Develop further an open, rule-based, predictable, non-discriminatory trading and financial system Deal comprehensively with developing countries' debt In co-operation with developing countries, develop and implement strategies for decent and productive work for youth In co-operation with pharmaceutical companies, provide access to affordable essential drugs in developing countries In co-operation with the private sector, make available the benefits of new technologies, especially information and communications.

health care, lack the amenities of safe drinking water and sanitation, cannot afford education for some or all of their children, and perhaps lack rudimentary shelter.'

The United Nations use another term, 'ultra poverty', to describe extreme poverty, defining it on a scientific basis as being when a household cannot meet 80 per cent of its minimum calorie requirements, even when using 80 per cent of its income to buy food (UNDP, 1997). This type of poverty occurs only in developing countries. A traditional indicator of extreme poverty that has been used extensively is the one developed by the World Bank, defined as people living on an income of less than US$1 per day per person (Sachs, 2005). The other common statistical measure of poverty used by the World Bank is an income of less than US$2 per day per person, which indicates moderate poverty (both the one-dollar and two-dollar indicators are measured at purchasing power parity).

One advantage of establishing these measures or 'poverty lines' is that they distinguish the poor from the non-poor, raise public awareness of poverty, and inform debate (Lanjouw, 1999). They subsequently permit a deeper analysis of the socio-economic and cultural characteristics of those in poverty and facilitate a longitudinal assessment of how their position changes through set periods of time. However, this approach assumes a clear relationship between poverty of income and disadvantage. This may not necessarily be the case, as income-based poverty calculations cannot account for cultural practices that may generate an inequality in livelihood options, e.g. opportunities for women may be suppressed by men (McMichael, 2004).

Based upon the World Bank's measure of income of US$2 per day per person, in the first decade of the twenty-first century, almost 50 per cent of the world's population live in poverty (World Bank, 2007). However, as is indicated by the earlier example of the television set, how poverty is conceptualised may rest on a more comparative or 'relative' basis. The concept of relative poverty is a common approach taken by researchers; typically it means having insufficient resources to meet socially recognised needs and to participate in wider society (Lister, 2004). The inability to participate in wider society may take the form of an inability to join in activities and to enjoy the living standards that are customary in the societies to which people belong (Townsend, 1979). Consequently, people may feel that they are excluded from the mainstream of society and denied opportunities to fulfil their own potential. Thus as Lister

(2004) suggests, poverty is not just a disadvantaged and insecure economic condition, but incorporates non-material aspects, including: disrespect; humiliation and low self-esteem; shame and stigma; lack of voice and powerlessness; and diminished citizenship.

The denial of opportunities for people to live a tolerable life and to realise their potential is a theme that has been developed by the influential and Nobel Prize-winning economist Amartya Sen. In Sens' analysis of poverty, the importance of income is not that it matters in its own right, but it is instrumental to what really matters, i.e. the kind of life that person is able to lead and the choices and opportunities open to them (Lister, 2004). Consequently, poverty can be viewed as an 'inadequacy' of income vis-à-vis a 'low' income (Sen, 1992). This means that poverty lines, e.g. people living on less than US$2 per day, may not be definitive in identifying those who are or are not in poverty. Basic human needs – such as avoiding a premature death, avoiding sickness that is preventable, having adequate nourishment, and having shelter and clothing – are key indicators of poverty alongside income level. More complex achievements would include taking part in community activities, leading a happy and stimulating life, and respect for oneself and for others (Sen, 1999). The relevance of this approach to understanding poverty was illustrated in the example of the Adivasi people, presented in Chapter 3. Although the Adivasi were techanically poor according to income measures, they would not have been willing to swap their lives for those of the materially much wealthier Germans, because they felt that the German lifestyle seemed alienated and lacked a sense of community.

Closely linked to having opportunities to improve one's life are 'capabilities' (Sen, 1999), which denote what a person 'can' do or be, i.e. the range of choices open to them. The concept of freedoms is also emphasised by Sen, including political, educational and personal freedoms. These, he believes, are determinatory for an individual's access to resources from which they may realise their capabilities. Without these freedoms, they may face a form of 'social exclusion', a term created by the sociologist Townsend in the 1970s. In the case of tourism, social exclusion might be experienced as an inability to have any participation in its economic benefits. For example, hawkers of handicrafts in developing countries may find that they are denied access to tourists, as a consequence of the tourist's movements being controlled by the established tourism industry, e.g. from a hotel enclave or cruise ship to a visitor attraction.

Local people may also find that, because of tourism development, they are excluded from access to natural resources that are necessary for their livelihoods, for example, beaches to lay out their fishing nets or water courses to irrigate their crops. Sometimes this may lead to conflict between stakeholders, as was described in Boxes 3.6 and 3.7.

The human poverty and development indices published in the annual UNDP Human Development Index (HDI) reflect Sen's approach (Lister, 2005). Built into the HDI are indicators that reflect three basic dimensions of human development: (i) a long and healthy life (measured by life expectancy); (ii) being educated (measured by adult literacy and enrolment at the primary, secondary and tertiary levels); and (iii) a decent standard of living (measured by purchasing power parity) (UNDP, 2006).

The relevance of being aware of these various types of poverty is that they require different strategic responses to tackle them. Thus, poverty is about more than purely being able to meet one's most basic needs for subsistence and survival. It also encompasses a denial of freedoms and capabilities to function in life and realise one's potential. Subsequently, whilst income may be an indicator of poverty, so are people's freedoms and opportunities to fulfil their potential. Such opportunities would include access to education, health care and democracy. So whilst poverty everywhere involves material deprivation, it also involves life-chances which will vary by culture and subculture. On this basis, human rights, dignity, security and participation in political processes have been integrated into the concept of poverty.

THINK POINT

How would you explain what poverty is?

Tourism and poverty

As was discussed in the introduction, tourism presents to developing countries a potential opportunity to alleviate poverty. An observable trend in the last few decades has been for increased tourism flows from the economically 'richer' countries of the world (i.e. the 'developed') to economically poorer ('developing'). A consequence of this trend has been an increase in tourism receipts in developing countries, both actual and as a percentage share of total world receipts, as shown in Box 5.2.

Referring to Box 5.2, in 1990 the developing countries had approximately an 18 per cent share of international tourism receipts, which by 2005 had risen to 30 per cent. The market share of the 50 least developed countries in the world doubled, rising from 0.4 per cent to 0.8 per cent. Whilst this 0.4 per cent rise may seem insignificant, it represents an actual increase of approximately US$4 billion dollars over 1990, although its equivalent purchasing power deflated to 1990 rates would be less than this figure. However, the potential economic significance of income flows at this level to the least developed countries is high, given their relatively low levels of economic development and relatively small GDPs.

The economic opportunities for countries in receipt of increased levels of international tourism receipts has led to an emphasis in tourism policy upon realising macro-economic benefits, including foreign exchange earnings and employment generation (as explained in Chapter 4) but without a specific focus upon the needs of the poor. Even within the remit of alternative tourism, the emphasis on ecotourism and community tourism has been upon the need to ensure that tourism does not erode the environmental and cultural base, rather than directly aiding the poor (Ashley et al., 2001). This latter point is particularly important, for if tourism is to be of use in combating poverty, the primary emphasis has to be upon targeting poor people.

Box 5.2

Developing countries' increasing share of international tourism receipts

	World		Developing countries		50 least developed countries	
Receipts	1990	2005	1990	2005	1990	2005
Actual (US$ billion)	273	682	50	205	1.1	5.3
Market share	100	100	18.1	30.1	0.4	0.8
% of international tourism receipts						

Source: UNWTO (2006b)

A notable advantage of tourism compared to other types of economic activity for combating poverty is that as the consumer travels to a destination, there exists an opportunity for a direct economic linkage between the tourist and the poor, for example, through the selling and buying of handicrafts. Importantly, this offers a chance not only for individuals to increase their incomes, but also to enhance the livelihoods of those in their family, who are also likely to be experiencing poverty (Ashley *et al.*, 2001).

Other arguments for tourism's use in the combating of poverty have their basis in the macro-economic benefits that may be gained through tourism development. For example, extra demand from tourism creates not only employment opportunities in tourism enterprises, but also potential ones in the organisations that supply goods and services to the tourism industry. However, the tourism sector in developing countries has a mixed record of employment opportunities for local people. Nevertheless, small-scale enterprise development aided by micro-finance schemes has a significant potential for aiding livelihood opportunities, as detailed in Box 5.3, which concerns the Makikwe Nature Reserve in South Africa. Importantly, gender-specific tourism employment opportunities for women may be particularly significant in combating poverty.

Significantly for achieving goal 3 of the MDGs, 'promote gender equality and empower women', tourism can offer women the opportunity for economic autonomy and greater influence in household decision-making (Scheyvens, 2002). Besides giving financial independence to women, employment in tourism may increase confidence and self-esteem, allowing women to have a more influential social and political role in society. However, as Kinnaird *et al.* (1994) observe, tourism operates within the wider economic, cultural and political contexts, all of which influence how opportunities for women will manifest themselves. In many cases, the employment opportunities for women have been more of the manual than the meaningful variety. According to Scheyvens (2002), the literature of women's experiences of employment in the tourism industry is overwhelmingly negative, with women occupying the majority of low-skilled and low-waged jobs. For example, in the Caribbean, the main form of employment for women in the tourism industry is as maids in hotels, which can be viewed as an extension of their domestic role and subsequently can be classified as unskilled and paid accordingly (Momsem, 1994).

Conversely, it can be argued that some types of tourism employment are attractive to women because of their time flexibility and demand for

Box 5.3

Madikwe Game Reserve, South Africa

Madikwe Game Reserve covers an area of 55,000 hectares and is located in South Africa's North West province. Previous to the reserves establishment in 1991, the region consisted of degraded cattle farms that had been expropriated from white farmers. In turn the establishment of the cattle farms had displaced tribal groups from their lands and most of the wildlife from the area had been hunted out.

The national government decided to turn the area into a game reserve, to be managed by the North West Parks Board, which enjoys an international reputation for people-based wildlife conservation. Alongside the objective to manage and restock the reserve with wildlife, the private sector was encouraged through leasehold agreements to invest in private lodges and infrastructure development to facilitate the growth of tourism. It was envisaged the benefits for local communities would include skills training, and that employment in tourism and income from the leasehold agreements would be used to fund community projects.

The project has been successful, introducing 8,000 mammals from 27 species into the reserve, including impala, wildebeest, elephants, zebras, buffalo, black and white rhino, lions, cheetahs, hyenas and the wild dog.

The project represents an economic model, in the sense that it was designed to generate a better income for the area, which in turn was reliant upon providing good opportunities for sighting game. Conservation was not the primary object of the project but it still had to be ecologically driven. The reserve has more than 30 tourist lodges, ranging from basic self-catering to seven-star operations. Critically in terms of community development, two of the three largest black communities around the reserve now own their own lodges, constructed on land that is leased from the Parks Board. The project has created more than 700 jobs for local communities, estimated to be five times more than if the area had been returned to agriculture.

The first lodge to open in the park was the Buffalo Ridge Guest House, which is owned by the Balete community. The funds to help build the lodge were supplied by the local government and the Ford Foundation, whilst the management of the lodge has been undertaken on a 15-year contract by the Nature Workshop, a Johannesburg management company. In return, they pay concession fees and give a percentage of their profits to the local community. They are responsible for training local people in tourism management, in conjunction with an on-going government lodge internship programme. The aim is that as employees become more skilled, they can gain higher positions in other lodges and new members of the community can then be trained. Keen to strengthen economic linkages with activities of the local community, the Balete use the money that they receive from tourism to resurrect community businesses that will service the lodges, e.g. brick-making.

Source: Grylls (2006)

cultural norms of hospitality; this makes it convenient to fit family responsibilities around them. For example, in many island communities throughout the Mediterranean region, it is often women who are the first to seize the financial opportunities of tourism, as an activity where their traditional areas of competence can be used (Scott, 2001). Also in Barbados, women beach vendors are generally older than their male counterparts, and find the job can be flexibly fitted in with other domestic duties (Momsem, 1994).

The tax revenues that can be gained from tourism offer opportunities to invest in education and health services to improve the livelihoods of the poor and to fund poverty reduction programmes. Other less tangible benefits of the poor being involved in tourism include psychological benefits and the enhancement of self-esteem. The range of benefits that tourism offers in combating poverty are summarised in Figure 5.2.

The use of tourism as a strategy for combating poverty has been encapsulated in the term pro-poor tourism (PPT), which is defined by Ashley *et al.* (2001: 2) thus: 'Pro-poor tourism is defined as a tourism that generates net benefits for the poor. Benefits may be economic, but they may also be social, environmental or cultural.' However, PPT is not a specific product or sector of tourism but rather an approach to securing an increase of the net benefits for the poor from tourism, and ensuring that tourism growth contributes to poverty reduction. It consequently enhances linkages between tourism businesses and poor people. Any type of company can be involved in pro-poor tourism, be it a small lodge, an urban hotel, a tour operator, an infrastructure developer. It also encompasses any type of tourism, for example, mass tourism or ecotourism, as long as it aims to provide benefits for the poor.

Although pro-poor tourism shares characteristics with sustainable tourism (ST), ecotourism (ET) and community-based tourism (CBT), key differences in focus exist between them. Whilst to varying degrees they all challenge the existing political economy of tourism – including characteristics of wealth distribution and intra-generational equity, democratic participation, the rights of women, and natural resource conservation – yet PPT is the only approach to have poverty as its key focus. The other significant difference between these types of alternative tourism is their geographical focus. PPT focuses exclusively on developing countries, whilst the other types may have application in a variety of geographical locations (Ashley *et al.*, 2001).

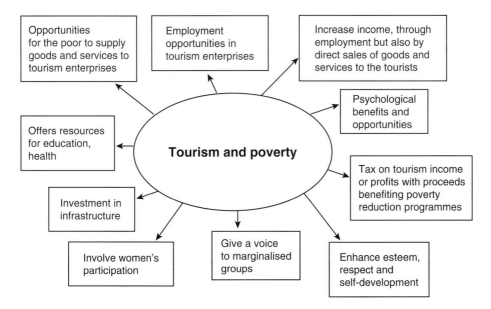

Figure 5.2 *How tourism can aid the reduction of poverty*

The other main strategy of the use of tourism to combat poverty has been from the UNWTO's Sustainable Tourism for the Elimination of Poverty (STEP) initiative. This initiative is explicitly linked to sustainable tourism development, having been launched at the World Summit for Sustainable Development in Johannesburg in 2002. Its intention is to 'creatively develop sustainable tourism as a force for poverty alleviation' (UNWTO, 2007a: 2). It has a direct objective of helping to achieve the MDGs, as summarised by Kofi Annan, the United Nations Secretary General:

> STEP will promote socially, economically and ecologically sustainable tourism, aimed at alleviating poverty and bringing jobs to people in developing countries … these objectives are fully consistent with the goals set out in the Millennium Declaration.

The STEP programme has a global perspective on the use of tourism in poverty alleviation. Five major areas have been identified for global action:

1 For developed countries to formulate pro-development strategies to encourage tourism to the world's poorest countries to enhance economic well-being, social development and mutual understanding;

2 For less developed countries to recognise the economic potential of tourism and make it a central focus of their Poverty Reduction Strategy Programmes;

3 For all countries to help poor countries to use tourism services to fight poverty and promote sustainable development;

4 To build a pro-development element into tourism strategies and to recognize the use of tourism in building understanding between people and enhancing global security;

5 For International Development Agencies to place tourism amongst their key priorities for infrastructure and entrepreneurial support; and for all tourism stakeholders to embrace the MDGs and pursue sustainable and responsible practices (UNWTO, 2007b).

Considerations and limitations of the use of tourism in combating poverty

Although PPT and STEP are significant policy initiatives, the utilisation of tourism to combat poverty is not without problems, which has led some to contest the view that tourism can actually alleviate poverty and assist poor people to have improved livelihoods. The basic tenet of tackling poverty through tourism rests upon ensuring the involvement of the poor as an economic activity. Consequently, it is reliant upon enlightened political leadership, philanthropy, resources and the ability to establish meaningful partnerships between different stakeholders. As the major influence on how the benefits of tourism are dispersed, the willingness of the private sector to adopt an ethos of providing opportunities for the poor will be critical to the success of PPT and STEP initiatives.

Although governments have the power to influence the tourism industry, they are unable to influence private investors to adopt non-commercial benefits for the poor. Nevertheless, the success of any poverty reduction strategy is dependent upon enlightened leadership by government. Critically, governments must recognise the potential of tourism as a means of alleviating poverty. Similarly, international donor agencies such as the World Bank and the Asian and African Development Banks must understand and be willing to support poverty reduction strategies through tourism.

Government and international development agencies also play a key role in attempting to establish meaningful links and co-operation between the established industry and the poor in attempts to give them access to participation in tourism. In meeting this objective, policy frameworks determined by governments are of major importance. This would include the allocation of resources for training, micro-loans for enterprise development, and fair-trade strategies. Examples of such initiatives undertaken by the Netherlands Development Organisation (SNV) and United Nations Development Agency (UNDP) are presented in Boxes 5.4 and 5.5.

Yet, tourism cannot have universal geographical application in combating poverty. An evident pre-requisite for tourism development is that there is an asset base of natural and cultural resources that is attractive for tourists. Consequently, tourism's use in combating poverty will be restricted to specific regions and locations, and subsequently not all

---

Box 5.4

Pro-poor tourism in Humla, Nepal

Humla is one of the most remote and under-developed regions of Nepal. Situated in the north-west corner of the country on the border with Tibet, in a composite index of development it is ranked the fourth poorest in Nepal. The people in the area suffer from severe food deficits and occasional disease epidemics. Development options are severely limited by its cold mountainous terrain (only 1 per cent of the land is available for agriculture) and poor infrastructure development (for example, the district capital of Simikot is 10 days' walk from the nearest road). Lack of opportunities for women is an evident issue at Humla as it is the lowest-ranking region of Nepal in terms of women's empowerment. The population of the district is approximately 50,000 people.

The Netherlands Development Organisation (SNV) has been active in improving trails and foot suspension bridges within the area. Besides aiding the movement of people and facilitating the movement of goods by animals (e.g. mules, yaks and yak-cow crosses alongside sheep, the traditional pack animal of the region), the main Hilsa–Simkot trail is now of a standard to permit trekking tourism. However, the numbers of tourists trekking on the route are in hundreds per annum, meaning that tourism accounts for only a small proportion of economic and social growth in the region. Much of the revenue from the limited amount of tourism rests with outside trekking agencies based in Kathmandu who sell the trips to the trekkers.

The majority of the people live below the international poverty line on an income of US$1 per day per person and very few have regular paid jobs. Some people are

Continued

Box 5.4 continued

'landless', struggling to produce even one month's worth of food. A problem in developing stronger tourism linkages within the area is the limited availability of local products and services for the outside agencies to use; for example, many of the Kathmandu trekking agencies bring all their food with them from Kathmandu. It is subsequently proposed to develop a multiple-use visitor centre for the region, where different tourism stakeholders could meet and exchange services, products and information. This would permit the co-ordination of services such as transportation, guides, portering equipment and agricultural produce, e.g. vegetables, fruit and poultry. The aim of this would be to allow the poor to have access to the tourism market. However, one of the problems faced in developing meaningful links between the private sector agencies and the poor is a cultural/political problem. Trekking companies have already established relationships with the local 'elite' who operate a monopoly on trekking work in the area. Subsequently, they are able to prevent competitors from participating in the market if they wish.

The role of the SNV is to act as facilitator for pro-poor tourism through the District Partners Programme (DPP), which aims to provide an institutional environment for sustainable economic development initiatives for women and men. This involves acting as a co-ordinator of stakeholders, including village committees, the private sector and NGOs who are working in the area. Emphasis is placed on participatory planning and involvement and capacity building for the poor to work in the tourism sector. This will include aspects of product development, marketing strategies, and the establishment of linkages with outside trekking agencies. Having identified the potential feasibility of tourism in Humla, training has now been given to the District Development Committees and NGO staff in sustainable and pro-poor tourism. The sourcing of funding for micro-tourism enterprise development is also another important role for SNV.

The practical outcomes of the project to date include:
- improved sanitation (over 400 toilets have been built on the trail);
- community support funds for the development of micro-enterprises have been approved;
- a tax of US$2 per tourist is now being levied;
- one community campsite has been developed.

Ultimately the success of the scheme depends upon political will, stakeholder partnerships and available resources. The vulnerability of tourism to external factors has been demonstrated by the civil war between the Maoists and the constitutional government, which has led to decreased tourism demand in Nepal. Hopefully, this situation now seems to have been resolved.

After Saville (2001)

Box 5.5

Nepal: tourism for rural poverty alleviation (TRPA) project

This project was funded by the United Nations Development Project (UNDP). Its aim was to tackle poverty through the creation of a sustainable tourism development project in line with Nepal's Ninth National Development Plan (NNDP) for the period 2000–5. It represented a holistic approach, which integrated objectives of other economic sectors for achieving poverty alleviation including health, transport, agriculture and environment. The primary aim of this approach was to demonstrate how tourism could be used for poverty alleviation and at the same time complement ministries' objectives. For example, an objective of the Ministry of Health was to provide clean drinking water to all communities and improve the levels of sanitation. This objective would need to be met through the TRPA project, because without a water supply and a suitable level of sanitation, sustainable tourism would not be possible. Concerning natural resources, an objective of the NNDP was to improve the biophysical environment of Nepal, and more specifically to remedy the environmental problems that arise from the existing economic and social conditions, especially poverty. The TRPA was seen to play a significant role in achieving this objective.

A range of pilot projects was established in impoverished districts of Nepal in 2001. The basis for identifying the impoverished districts was a series of indicators developed by the United Nations Development Program. These included life expectancy and health standards; literacy levels; remoteness and lack of public infrastructure; gender inequality; and per capita income. Six districts were chosen, from which target groups were identified to maximise the impacts tourism could have on alleviating poverty and aiding the poor. By 2006, tourism ventures had been established with 48 Village Development Committees (VDCs) in impoverished parts of Nepal.

Inherent to the approach of TRPA was the participation of local communities in tourism ventures. Thus emphasis was placed upon impoverished village communities undertaking self-assessment in terms of their capabilities and resources for tourism. The extent to which the scheme has been successful is unknown as it awaits detailed assessment. However, it has brought together three levels of government, i.e. national agencies and ministries, District Development Committees, and VDCs, with industry stakeholders and community empowerment. Examples of practical projects include the construction of a Sherpa porters' shelter at Namche for 60 porters with beds, toilets, washing facilities and a kitchen. Prior to this there were no facilities for porters, and death and injuries to porters is a major ethical issue of trekking tourism in Nepal. Another project was the establishment of a micro-hydro energy project in Phortse village in Lower Sagamartha. This is an area that had substantial tourism potential but had suffered a major energy crisis resulting in deforestation. A further initiative was capacity-building workshops, including environment and waste management, conservation, lodge management and accountancy.

Source: Rossetto, *et al.* (2007)

developing countries will be able to benefit equally. Additionally, as Ashley *et al.* (2001) comment, whilst poverty reduction through PPT can be significant at a local or district level, its national impacts would vary with location and the relative economic importance of tourism.

The characteristics and hegemonic relationships of the existing political economy will also be influential in determining the success of tourism and poverty initiatives. Local tenure and land ownership issues will be important in determining not only how land is used for tourism but also the extent to which poor people have access to resources for their livelihoods. Where poor people have no land ownership rights, they are vulnerable to development decisions which may act directly against their own interests. In contrast, tenure over land, wildlife or other tourism assets can give the poor market power, enabling them to play a participatory part in decision-making and secure benefits from tourism (ibid.). Similarly, the political, educational and personal freedoms highlighted by Sen (1999) will be critical for poor people's ability to improve their livelihoods through participation in tourism.

Partnerships between the different tourism stakeholders will be essential to the success of PPT schemes. Besides an enlightened approach from national governments, all the stakeholders in tourism will need to adopt an informed and philanthropic approach to how tourism can be used to help the poor. Issues of ethics and working for a greater social good vis-à-vis an emphasis on individual benefit will be determinatory in the success of any poverty strategy. For example, hotels and tour operators need to work with local communities and local government to develop a model of tourism that actively presents them with livelihood opportunities.

Similarly, marketing strategies need to attract tourists who are supportive to enhancing the livelihoods of the poor, particularly those who like to visit markets and pursue tourism experiences based upon nature, culture and everyday life that are likely to be provided by the poor people (UNWTO, 2007b). This can be enhanced by attracting domestic tourists, who are more likely to share the cultural values and practices of local people. The use of organised markets in prime locations can greatly facilitate local sales to tourists (Ashley *et al.*, 2001). For example, women craft sellers have sites within some parks in KwaZulu Natal in South Africa, while at Gonarezhou National Park in Zimbabwe, one of the demands of local communities is for a market at the park entrance to access incoming tourists to the park.

A further key tenet of poverty reduction is the strengthening of inter-sectoral linkages with tourism to enhance economic benefits for the poor. One sector that would benefit particularly from this approach is agriculture. As Torres and Momsem (2004) note, agriculture remains the livelihood of most of the poorest people in developing countries. Tourism has the potential to create extra demand for local agricultural produce. However, the development of this backward linkage is dependent upon the tourism industry being willing to facilitate communications and negotiations with local farmers, and upon the farmers' ability to supply produce of a quality and regularity that is demanded by the industry.

A further reason for the necessity to develop linkages to other economic sectors such as agriculture is that without it, there is a danger that a polarised form of development based upon tourism may take place (Brohman, 1996). In this scenario, as tourism becomes more successful, the rest of the economy fails to follow, causing a disparity of wealth between tourism and other types of tourism activity (Torres and Momsen, 2004). The medium- to long-term effects of this are for more natural and human resources to be used for tourism as it becomes more financially beneficial than other economic sectors. The ultimate effect may be to create an economic over-dependency upon it and a danger of loss of livelihoods if tourism demand falls.

The threat of a decrease in tourism demand, combined with the vulnerability of tourism demand to external factors, is a significant criticism of the use of tourism as means for development or tackling poverty. Events such as terrorism, natural disasters, economic recession and changing market tastes all threaten tourism demand. Although these may be short term, there is nevertheless a danger of over-emphasis or overuse of tourism as a means of poverty reduction, necessitating that is balanced with other types of development.

Summary

- There is an explicit link between poverty and environmental degradation. Whilst the focus of tourism has been upon development, since the 1990s there has been an increased emphasis on how tourism can be used to target the poor and reduce poverty. The use of tourism has a special relevance in regions of developing countries and the least developed countries to help achieve the MDGs.

- The meaning of 'poverty' is not fixed and may be politically contentious. There are various types of poverty, including: absolute; extreme; ultra; and comparative. Subsequently, poverty is subject to different interpretations. Poverty lines are sometimes used as a benchmark, the most common being the ones established by the World Bank of an income of less than US$1 per day per person (extreme poverty) and of US$2 per day per person (moderate poverty). Another way of conceptualising poverty is to think of it as a denial of freedoms, e.g. the freedom to work to earn an income; access to education, health care and democratic participation.

- For tourism to combat poverty, it needs to be targeted to provide economic opportunities for the poor. These offer the potential not only to increase individuals' income but also to enhance the livelihoods of their families. Small-scale opportunities aided by micro-finance schemes have a significant potential for aiding livelihood opportunities. The employment opportunities created for women may be significant in combating poverty.

- A significant strategic initiative in using tourism to combat poverty is pro-poor tourism (PPT), which originated from the United Kingdom's Department for International Development (DFID) in the late 1990s. Its aim is to use tourism to generate economic, social, environmental and cultural benefits for the poor. PPT is not a specific type of tourism but a philosophy to ensure that the benefits of tourism for the poor are increased and that tourism growth contributes to poverty reduction. Its geographical focus is exclusively on developing countries. At the World Summit for Sustainable Development in 2002, the UNWTO launched their Sustainable Tourism for the Elimination of Poverty (STEP) initiative. The emphasis of this initiative is to use sustainable tourism as a force for poverty alleviation.

- However, the utilisation of tourism to alleviate poverty is not unproblematic. It is reliant upon enlightened leadership and philanthropy from tourism stakeholders. Critically, it is essential that the private sector actively support PPT or STEP programmes. The geographical scope of tourism and its subsequent use in alleviating poverty is also restricted to areas with suitable natural and cultural resources. Issues of political economy, including land ownership rights and access to resources, are also critical to the success of tourism poverty initiatives. Tourism demand is vulnerable to external factors, e.g. terrorism, natural disasters and economic recession. Consequently, it is necessary to ensure that an over-dependency is not placed upon tourism as a means of alleviating poverty.

Suggested reading

Ashley, C., Boyd, C. and Goodwin, H. (2000) *Pro-Poor Tourism: Putting Poverty at the Heart of the Tourism Agenda*, London: Overseas Development Institute.

Harrison, D. (ed.) (2001) *Tourism and the Less Developed World: Issues and Case Studies*, Wallingford: CAB International.

Lister, R. (2004) *Poverty*, London: Routledge.

Scheyvens, R. (2002) *Tourism and Development: Empowering Communities*, Harlow: Pearson Education.

Sharpley, R. and Telfer, D. (eds) (2002) *Tourism and Development: Concepts and Issues*, Clevedon: Channel View Publications.

Suggested websites

Institute of Development Studies: www.ids.ac.uk/ids

Overseas Development Institute: www.odi.org.uk

Pro-poor tourism: www.propoortourism.org

United Nations: www.un.org

United Nations World Tourism Organisation: www.world-tourism.org

The World Bank: www.worldbank.org

6 Sustainability and tourism

- Origins of sustainable development
- Meaning of sustainable development
- Perspectives on its interpretation
- Its application to tourism

Introduction

Concern over the negative effects of development upon the environment has led to calls for a new conceptual approach to development. This conceptual approach is termed 'sustainable development', and it has become a new paradigm for all forms of development, including tourism.

The origins of 'sustainable development'

In contemporary terms, the term 'sustainable development' is usually credited to the Brundtland Report, officially the report of the World Commission on Environment and Development (WCED, 1987). The origins of 'sustainability' as opposed to 'sustainable development' lie in concerns over conservation and can be traced back to the conservation movement of the mid-nineteenth century (Stabler and Goodall, 1996). The concept of 'sustainable development' first originated in the World Conservation Strategy published by the World Conservation Unit (IUCN) in 1980 (Reid, 1995).

However, the popularisation of the term did not occur until its use in the Brundtland Report seven years later, perhaps because by 1987, environmental awareness was at a much higher level. The report was based upon an inquiry into the state of the earth's environment, led by Gro Harlem Brundtland, the Norwegian Prime Minister, at the request of the General Assembly of the United Nations. Concern over the effects of the pace of economic growth on the environment since the 1950s led the United Nations in 1984 to commission an independent group of 22 people from various member states representing both the developing and developed world, to identify long-term environmental strategies for the international community (Elliott, 1994). The key environmental concerns of the United Nations were the high levels of unsustainable resource usage associated with development, and the role of pollution in major environmental problems such as global warming and depletion of the ozone layer, which threatened human well-being.

Accompanying the heightened awareness of environmental problems was also a realisation that the environment and development are inexorably linked. Development cannot take place upon a deteriorating environmental resource base; neither can the environment be protected when development excludes the costs of its destruction, as outlined in Chapter 4. Although 'development' is a common term in today's language, it is only since the 1950s that development has been studied as an academic subject, the time at which colonial territories started to achieve independence (ibid.).

THINK POINT

Growth and development are not the same. Would you characterise the tourist boom along the Spanish costas as an example of growth or development? Give your reasons.

'Development' and 'growth' are often used as synonymous terms but there is a critical difference between the two. Growth means to get bigger or larger, whilst development refers to a change in state, for the better.

The pace of economic growth since the 1950s has been rapid; for example, although industrial production grew 50-fold in the twentieth century, 80 per cent of this growth took place since 1950 (WCED, 1987). Whilst this rapid pace of industrial growth has helped to increase the living standards and life expectancy for many, the number of people in the world who continue to live in absolute poverty (i.e. having insufficient income to meet the basic needs of food, shelter and clothing) is approximately 1 billion people, or 20 per cent of the world's population (Reid, 1995). The way development has been pursued, characterised by a general lack of concern for the environment, has led to the use of natural resources in a way that is unsustainable; that is, many finite resources are being exhausted whilst the capacity of the natural environment to assimilate waste is being exceeded.

It was the predominance of the negative aspects of development changes that led to the calls for sustainable development. The term gained greater attention following the United Nations Conference on Environment and Development (UNCED), held in Rio de Janeiro in June 1992, popularly referred to as the 'Earth Summit'. At the Earth Summit a programme for promoting sustainable development throughout the world, known as Agenda 21, was adopted by participating countries. Agenda 21 is an action plan laying out the basic principles required to progress towards sustainability. It envisages national sustainable development strategies, involving local communities and people in a 'bottom–up' approach to development, rather than the 'top–down' approach which has typically characterised national development plans.

Although tourism as an economic sector was not debated in Rio, five years later at the 'Earth Summit II' in New York, it was debated as a recognised economic sector. In the recommendations and outcomes of the report it was stated:

> The expected growth in the tourism sector and the increasing reliance of many developing countries, including small island developing States, on this sector as a major employer and contributor to local, national, sub regional and regional economies highlights the need for special attention to the relationship between environmental conservation and protection and sustainable tourism.
>
> (Osborn and Bigg, 1998: 169)

In the last decade of the twentieth century, the term 'sustainable development' became widely used by governments, international lending agencies, non-governmental organisations, the private sector and academia. As Farrell and Twining-Ward (2003: 275) observe, faced by enormous ecological change driven by human action, sustainable development 'has evolved over three decades from an environmental issue to a socio-political movement for beneficial social and economic change'. An encapsulated history of the concept of sustainable development is presented in Box 6.1.

The meaning of sustainable development

The fact that the term 'sustainable development' can be adopted by governments, international lending agencies, non-governmental

Box 6.1

The origins of sustainable development

- The term 'sustainable development' can be traced back to the conservation movements of the mid-nineteenth century
- Established as a policy consideration in the World Conservation Strategy published by the World Conservation Union (IUCN) in 1980
- Term gains greater attention and popularity after the publication of the Brundtland Report (1987)
- The 'Earth Summit' 1992 held in Rio de Janeiro adopts 'Agenda 21', aimed at promoting sustainable development throughout the world
- Tourism is recognised as an economic sector that needs to develop sustainably at 'Earth Summit II' in 1997 in New York.

organisations, the private sector and academia, some of whom could be viewed as having divergent and politically opposed objectives, is a reflection of the inherent ambiguity of the concept. This ambiguity permits a variety of perspectives to be taken on sustainability. Much of this ambiguity can be traced back to the most commonly quoted definition of sustainable development taken from the Brundtland Report:

> Yet in the end, sustainable development is not a fixed state of harmony, but rather a process of change in which the exploitation of resources, the direction of the investments, the orientation of technological development, and institutional change are made consistent with future as well as present needs.
>
> (WCED, 1987: 9)

Richardson (1997) describes this definition as a political fudge, aimed at compromising the opposing views of commissioners from different states, to keep everyone happy. However, the remainder of the Brundtland Report makes it clear that within the scope of this definition other key issues relating to development have to be addressed, such as the alleviation of poverty, degradation of the environment, and issues of intra- and inter-generational equity.

It is important to realise that sustainable development is not concerned with the preservation of the physical environment but with its development based upon sustainable principles. The Brundtland Report stresses the need for the alleviation of global poverty, not only as an ethical objective, but also as a key method to ameliorate the pressures being placed upon the physical environment. Subsequently the report lays out the principles of how the environment should be developed:

> Economic growth and development obviously involve changes in the physical ecosystem. Every ecosystem everywhere cannot be preserved intact.... In general, renewable resources like forests and fish stocks need not be depleted provided the rate of use is within the limits of regeneration and natural growth.
>
> (WCED, 1987: 45)

Emphasis is therefore placed upon the conservation of the resource base rather than the preservation of individual flora and fauna. In terms of the use of finite resources, such as minerals and fossil fuels, actions such as resource management, recycling and economy of use are highlighted to make sure that the resources do not run out before substitutes are found.

A central theme of the Brundtland Report, poverty alleviation through sustainable development, is critical for the long-term environmental well-being of the planet. Poverty is a major cause of environmental destruction as explained in the last chapter, a relationship that is particularly exacerbated in regions of the world where the population is growing rapidly, and forced into more marginal environments. The link between poverty and environmental destruction is emphasised by Elliott (1994: 1) in the following passage:

> In the developing world, conditions such as rising poverty and mounting debt form the context in which individuals struggle to meet their basic needs for survival and nations wrestle to provide for their population. The outcome is often the destruction of the very resources with which such needs will have to be met in the future.

Poverty has been worsened in many areas by population pressure, and the growth in world population in the twentieth century was rapid. In 1900 the world's population was 1,600 million, by the beginning of the new millennium it has risen to 6,000 million, and according to the UN by the year 2025 it is estimated it will be over 8,500 million (World Guide, 1997).

Different perspectives on sustainable development

That there should be differing perspectives on the meaning of sustainability is perhaps unsurprising, given the range of priorities, interests, beliefs and philosophies underpinning human interaction with the environment. Two broad ideological approaches to the environment are recognisable: 'technocentrism' and 'ecocentrism'. *Technocentrism* is characterised by a belief that technical solutions can be found to environmental problems through the application of science, thereby placing its faith in quantifiable solutions to problems. Its reliance upon quantification allows for a divorced objectivity in decision-making, rendering subjective considerations over the environment, such as feelings or emotions, as being unworthy of consideration. Such a reliance upon objective measurement means that the environment's complexity as a system can be overlooked and differing viewpoints ignored. As Reid (1995: 131) remarks:

> Technocentrism's readiness to quantify problems and solutions and its forecasts of success make it attractive to decision-makers who may be ignorant of the complexities of issues filtered through viewpoints, interpretations and value judgements of which they may be unaware

Through a lack of appreciation of the complexity of the environment as a system, there is also a lack of realisation that most development decisions have consequences for the environment that extend beyond the physical boundaries of an individual project. Within the philosophy of technocentrism, the physical environment is viewed as a resource to be exploited by humans as they deem appropriate. As for the degree of democracy in decision-making, technocentrism is characterised by a preference for centralised control rather than local decision-making, as Pepper (1993: 34) writes: 'There is little desire for genuine public participation in decision making, especially to the right of this ideology, or for debates about values.'

An alternative ideological approach to how we view the environment is *ecocentrism* (O'Riordan, 1981). Ecocentrism is closely associated with the philosophies of the romantic transcendentalists and characterised by a belief in the wonderment of nature. Ecocentrics have a lack of faith in both modern technology and technical and bureaucratic elites, and subsequently advocate alternative technologies as a way forward. This is not only because alternative technologies are likely to be more environmentally benign, but also because they are more democratic in the sense that they can be owned, maintained and understood by individuals with little economic or political power. Ecocentrics' position on technology may be termed 'Luddite'; that is, they are not opposed to new technology, but are opposed to technology that places its ownership and control in the hands of a powerful elite.

Doyle and McEachern (1998) equate the term 'ecocentrism' with 'deep ecology', recognising that it is based on four main premises:

1 That all beings, whether human or non-human, possess an intrinsic value, the antithesis of the technocentric instrumental viewpoint of nature
2 That all beings are of equal value and there is therefore no hierarchy of species in nature
3 That all nature is interconnected, with no dividing lines between the living and the non-living, the animate and inanimate, or the human and non-human
4 That the earth is finite in its carrying capacity.

These views are in contrast to the dominant world-view on development, which incorporates a pioneering and frontier mentality, assuming an unlimited supply of natural resources and an unlimited waste absorption capability in nature. The different approaches to development between the 'dominant world-view' and 'deep ecology' are shown in Box 6.2.

The equating of the dominant world-view with the attitudes of technocentrism is not difficult and, according to O'Riordan (1981: 1), the balance of power in decision-making lies with the technocentrics: 'Technocentrists tend to be politically influential for they usually move in the same circles as the politically and economically powerful, who are soothed by the confidence of technocentric ideology and impressed by its presumption of knowledge.'

Within these two broad ideological approaches, a range of contrasting positions can be taken on sustainable development. Baker *et al.* (1997) conceptualise a ladder of sustainable development moving upward from a technocentric approach at the bottom to an ecocentric position at the top. At the bottom of the ladder is what they term the 'treadmill' approach, focused on accruing material products and pursuing wealth creation. This approach is strongly technocentric in character, with little concern being displayed for the environment, beyond believing that

Box 6.2

Different approaches to development between the 'dominant world-view' and 'deep ecology'

Dominant world-view	*Deep ecology*
• Strong belief in technology for progress and solutions	• Favours low-scale technology that is self-reliant
• Natural world is valued as a resource rather than possessing intrinsic value	• Sense of wonder, reverence and moral obligation vis-à-vis the natural world
• Belief in ample resource reserves	• Recognises the 'rights' of nature are independent of humans
• Favours the objective and quantitive	• Recognises the subjective such as feelings and ethics
• Centralisation of power	• Favours local communities and localised decision-making
• Encourages consumerism	• Encourages the use of appropriate technology
	• Recognises that the earth's resources are limited

Source: After Bartelmus (1994)

human ingenuity can solve any environmental problem created by development. The second rung of the ladder is occupied by 'weak sustainable development', in which the aim 'is to integrate capitalist growth with environmental concerns' (ibid. 13). From this perspective emphasis is placed upon continued economic growth, but the environmental costs of development are to be reflected in economic calculations and accounting procedures. Unlike the 'treadmill' approach, 'weak sustainable development' recognises the finite nature of many natural resources, and also recognises that the environment's capacity to assimilate waste is not limitless. Emphasis is therefore placed upon furthering the development of economics as a discipline to permit a better evaluation of environmental assets. A criticism of this approach is that it limits the value of the environment to that which can be quantitatively measured, neglecting its cultural or spiritual worth.

The third rung of the ladder is occupied by an approach more synonymous with ecocentrism than with technocentrism. 'Strong sustainable development' advocates that environmental protection be a precondition of economic development. Consequently, within this perspective the environment becomes the key consideration, rather than economic growth as in the latter two scenarios. This perspective requires that development policies aim to maintain the productive capacity of environmental assets, and preserve other environmental assets deemed worthy of protection as they are, for example, tropical rainforests. 'Strong sustainable development' takes qualitative aspects of the environment into account and 'local communities' will be involved in decision-making over development issues. All the available instruments of policy, including legal, fiscal and economic measures, should be used and adapted to support this approach.

Occupying the top rung of the ladder is what Baker refers to as the 'Ideal Model'. This approach is underpinned by a strong ethical dimension, the viewpoint being taken that nature and non-human life have an intrinsic value, which extends beyond their usefulness to humans. The measurement of growth expressed in quantitative terms ceases to be relevant as the 'quality of life' becomes the objective of development rather than the 'standard of living'. The policy implications of this viewpoint are that environmental protection will place severe restraints on the use of the earth's resources and economic activities. This approach represents a 'radical' approach to sustainable development, involving structural changes in global society and economy. Indeed, according to Baker, ecologists would argue that the 'Ideal Model' represents a new paradigm in its own right.

The advocation of radical changes in society to achieve sustainable development concentrates on power relationships in the wider political economy. This entails addressing the root causes of non-sustainability, including the distribution of power and wealth, the roles of transnational corporations, class-based politics, and gender inequalities. The distribution of intra-generational equity, even though it was an issue raised in the Brundtland Report as having direct consequences for poverty, is notably absent from most government agendas. Radical approaches to sustainable development challenge the values and principles of capitalist society, as Doyle and McEachern (1998: 37) comment: 'Radical environmental political theorists are involved in paradigm struggles, each seeking to create new sets of key values and principles that directly challenge existing, powerful paradigms.'

Radical political positions over the interpretation of sustainable development, which emphasise different degrees and elements of structural change, include 'deep ecology', 'social ecology', 'eco-socialism' and 'eco-feminism'. The values of deep ecology in contrast to the values of the dominant world-view have already been illustrated in Box 6.2. To reiterate, the emphasis of deep ecology rests on the intrinsic value of all life, and acknowledging that all life is equal. This means that deep ecologists take the viewpoint that human beings are not the sole purpose of the evolutionary process but just another animal species. In this philosophy there is no theoretical separation between humans and the rest of nature. These beliefs have led to campaigns for the preservation of wilderness in North America, Scandinavia and Australia. Consequently, deep ecologists are sometimes enthusiastic supporters of population control programmes in order to reduce destructive human pressures upon the earth, which has led to accusations of them holding fascist tendencies.

The politics of *social ecology* are rooted in anarchist traditions, with their associated hostility to the state, liberals and Marxists (Doyle and McEachern, 1998). Social ecologists subsequently advocate maximum individual autonomy and the decentralisation of society into local communities, as opposed to the concept of the nation-state. According to Roussopoulos (1993: 122), social ecology: 'represents the greatest advance in twentieth century eco-philosophy'. Its development is closely associated with the American ecologist Murray Bookchin, who adopts two key positions in his philosophical approach to social ecology. The first is to recognise that the many types of hierarchical structures that exist in society are socially constructed and determined, and represent the source of all forms of domination within society, and also between humans and nature. The second position, which distinguishes social ecology from deep ecology, is that although humans are part of nature, they occupy a rather more exalted place in the evolutionary scheme (Doyle and McEachern, 1998).

Eco-socialists in contrast to social ecologists attempt to integrate ecological principles with the political theories of socialism including Marxism, rather than anarchism. The concern of eco-socialism lies equally with the problems of social justice, such as the inequity of wealth distribution in society, and ecological concerns. Just as there are many types of socialism, so there are many varieties of eco-socialism. Although eco-socialism can be allied to Marxism, the theories of Marx that assume a limitless abundance of nature are not accepted by eco-socialists

(Roussopoulos, 1993). A distinguishing feature between social ecologists and eco-socialists is their belief in the nation-state. Eco-socialists do not accept that environmental problems can be tackled at a localised level, favouring central government and the strengthening of pan-national bodies, like the United Nations. Consequently, a necessary precursor for environmental protection is for eco-socialists to have power.

According to Roussopoulos (1993), *eco-feminism* has its origins in anti-militarism and links the environmental problems that the world is facing today to the system of patriarchy. Therefore, ecological destruction and the oppression of women are linked. As for eco-socialism, there is a broad range of differing political perspectives, including cultural eco-feminism, liberal eco-feminism, social eco-feminism and socialist eco-feminism.

Political tension over sustainable development therefore underlies much of the debate about its interpretation. A fundamental division is between, on the one side, those for whom sustainability represents little more than improving technology and environmental accounting systems, whilst preserving the status quo of existing hierarchies and power structures in society and, on the other side, those who have more radical political agendas, involving changing the value systems and power structures of society.

Sustainability and tourism

In terms of the application of the concept of sustainable development to tourism, varying perspectives have been adopted, which are a reflection of the wider political debate about sustainable development outlined in the preceding section of this chapter. Since the early 1990s, the sustainable tourism debate has become more holistic to cover not just environmental issues but also socio-cultural, economic and political dimensions (Bramwell, 2007). At its most simplistic, a broad differentiation can be made between 'sustainable tourism', in which emphasis is placed on the customer and marketing considerations of tourism to sustain the tourism industry and 'sustainable development', in which emphasis is placed on developing tourism as a means to achieve wider social and environmental goals. Therefore sustainable tourism will not necessarily equate with the aims and objectives of sustainable development.

By the early 1990s, Coccossis and Parpairis (1996) and Hunter (1996) had identified a form of sustainable tourism oriented towards the viability of the tourism industry. Coccossis and Parpairis refer to this as the 'economic sustainability of tourism' and Hunter as the 'tourism imperative', the aim of which is primarily concerned with satisfying the needs of tourists and players in the industry. The central justification of this approach is that tourism development is to be encouraged and seen as acceptable, if developing other economic sectors, such as logging and mining, are viewed as being more environmentally damaging than tourism. However, this perspective fails to allow for the fact that the negative consequences of tourism development are incremental and cumulative, and that prejudgement of the likely environmental impacts of tourism development through techniques such as environmental impact analysis is therefore difficult, as is discussed in Chapter 7. Comparative impact analysis with other types of development alternatives therefore becomes meaningless and misleading.

Hunter's second scenario is when the environmental resources for tourism receive consideration, but are secondary to the growth of the tourism sector, an approach known as 'product-led tourism'. Hence although environmental and social concerns may be more important than in the former scenario, their relevance is still largely equated with maintaining the existing tourism product. Hunter suggests this is a 'weak' interpretation of sustainable development, but that such a position is justified in communities heavily dependent upon tourism, where placing a high priority on environmental concerns may mean that the well-being of the community is threatened. Development may have already led to extensive alteration of the natural environment and attention will therefore focus on environmental improvements to the existing development.

Hunter's third scenario is termed 'environmentally led tourism', in which types of tourism would be promoted that are reliant upon a high-quality environment. The main aim would be to make the link between the success of the tourism industry and environmental conservation so obvious to all the stakeholders that stewardship of the environment becomes a priority. This is analogous to what Coccossis refers to as 'sustainable tourism development', where the protection of the environment is seen as a key component of the long-term economic viability of the industry. The main difference compared to 'product-led tourism' is that the environment is prioritised and forms of tourism are developed that are not damaging to it. Tourism would therefore be centred upon attracting tourists who would wish to be educated about the natural environment and perhaps participate in its conservation.

The final scenario suggested by Hunter is 'neotenous tourism'. In some exceptionally sensitive areas, where the preservation of species is paramount because of their ecological significance, tourism should not be permitted at all. In other areas that are viewed as being ecologically significant, tourism should be limited to very small numbers through policy control measures.

The focus of both Coccossis and Parpairis's (1996) and Hunter's (1996) work is the physical environment. However, the concept of sustainability should be broadened to include cultural, political and economic dimensions. The ambiguity of the concept of sustainability means that the political context, and especially the political values of those who have power and decision-making, will be influential in determining the interpretation of sustainable tourism. Butler (1998) points out that it is not possible to separate sustainable tourism from the value systems of those involved and the societies in which they exist. As Mowforth and Munt (1998: 122) remark:

> If it remains a 'buzzword' which can be so widely interpreted that people of very different outlooks on a given issue can use it to support their cause, then it will suffer the same distortions to which older-established words such as 'freedom' and 'democracy' are subjected.

Mowforth and Munt contend that if sustainable tourism is really to be achieved, then there is a 'need to politicise the tourism industry in order to promote its movement towards sustainability and away from its tendency to dominate, corrupt and transform nature, culture and society' (1998: 123).

House (1997) also infers a political dimension to the application of sustainability in tourism. She recognises two polarised positions or 'schools of thought' reflecting different ideologies on the employment of the concept. At one end are the 'reformists', whose ideology and actions of implementation are preoccupied with the status quo, while 'structuralists' possess a much more radical view of tourism development, which challenges the paradigm on which economic, social and political development is based. Reformists are therefore 'reluctant to challenge the existing social, political and economic structures that underpin tourism development and behaviour' (ibid.: 93).

Reformists do not challenge the values in society that create environmental and social problems but try to manage them. House draws a distinction between those who modify existing models of tourism to make them more 'sustainable', and those who use tourism as a vehicle to experiment with 'alternative existence approaches' to achieve an alternative social order to the present one. The structuralist model of tourism is therefore one that is radical, involving questioning the values of society, as much as those of tourism. The relationship between the ideas of 'reformists' and 'structuralists' and sustainable tourism is analogous, with the differing positions taken on the meaning of sustainability within the field of political ecology, as outlined earlier in the chapter.

In a comprehensive review of how sustainability has manifested itself in tourism, Saarinen (2006) identifies three traditions. The first, the 'resource-based tradition', emphasises conservation and the need to 'protect' nature and culture from unacceptable changes by tourism activities. In this tradition there are conspicuous links to conservation management techniques, such as 'carrying capacity', explained in the next chapter. The second, 'activity-based' tradition, is centred upon an acceptance that tourism development can contribute to sustainability. This is a position that is strongly advocated by the tourism industry in a desire to sustain tourism and its resource base for future development, aiming to sustain the capital investment in tourism. The third 'community-based' tradition challenges the existing political economy by advocating the wider involvement of various stakeholders, especially host communities. A difference between the resource-based traditions vis-à-vis the other two is that in the first, sustainability is viewed as a physical and 'real' construct based upon conservation and measurement. In contrast, the activity-based and community-based traditions have a bias towards a social construction of sustainability, in which judgements are made about acceptable levels of trade-off between economic and social gains against natural resource losses. The key difference between these two is the hegemonic relationships between the stakeholders and those parties who hold the absolute power of decision-making. These two traditions also incorporate themes of political ecology, in the sense of considering the power relationships that exist in determining which stakeholders have access to and management of natural resources. However, as Bramwell (2007) observes, political ecology is an area of social studies that has been given insufficient consideration in sustainable tourism. Similarly, Liu (2003) observes that the issue of intra-generational equity referred to in the Brundtland Report has received scant attention in sustainable tourism.

It is therefore necessary to realise that sustainable tourism is not merely connected with conservation or preservation of the physical environment but incorporates cultural, economic and political dimensions. As has been pointed out, the ambiguity of the term means that it can be interpreted and owned by so many different groups with opposing ideologies that trying to agree on a common definition of the term is meaningless. Perhaps the most useful way of thinking about sustainability is not necessarily to think of it as an end-point, but rather as a guiding philosophy which incorporates certain principles concerning our interaction with the environment. The next section of this chapter considers the guiding principles that have been developed in connection with tourism.

> **THINK POINT**
>
> Does 'sustainable tourism' mean the same thing as 'sustainable development'?

Sustainability in action

The concept of sustainability has been applied in the tourism sector in different ways, at both national and local levels, and in the public and private sectors. The last decade of the twentieth century saw an increased effort by some private sector tourism organisations to make it evident that they were placing the environment in a more central position to their operations and attempting to become more 'sustainable', as discussed in Chapter 7. The extent to which this represents a genuine concern for the environment, or a business ploy to attract more customers and an attempt to stave off regulation of the industry, is uncertain. Butler (1998: 27) suggests that the tourism industry has adopted sustainability for three reasons: 'economics, public relations and marketing'.

One of the first public strategies on tourism and sustainability emerged from the Globe'90 conference in Canada, which brought together government, non-governmental organisations (NGOs), the tourism industry, and academics to discuss the future relationship of tourism with the environment. Five main goals of sustainable tourism were identified:

1 To develop greater awareness and understanding of the significant contributions that tourism can make to the environment and economy
2 To promote equity and development
3 To improve the quality of life of the host community

4 To provide a high quality of experience for the visitor
5 To maintain the quality of the environment on which the foregoing objectives depend (Fennell, 1999: 14).

As with the concept of sustainability, the goals tend to be all-encompassing, potentially conflicting and give little guidance on how tourism should be developed.

In Britain, the Department of the Environment (DOE) developed guiding principles for the development of tourism in the early 1990s (see Box 6.3).

The first of these principles is significant because of its recognition that the environment has an intrinsic value; that is, that nature has a

---

Box 6.3

Guiding principles of sustainable tourism

- The environment has an intrinsic value which outweighs its value as a tourism asset. Its enjoyment by future generations and its long-term survival must not be prejudiced by short-term considerations.
- Tourism should be recognised as a positive factor with the potential to benefit the community and the place as well as the visitor.
- The relationship between tourism and the environment must be managed so that the environment is sustainable over the long term. Tourism must not be allowed to damage the resource, prejudice its future enjoyment or bring unacceptable impacts.
- Tourism activities and development should respect the scale, nature and character of the place in which they are sited.
- In any location, harmony must be sought between the needs of the visitor, the place and the host community.
- In a dynamic world some change is inevitable and change can often be beneficial. Adaptation to change, however, should not be at the expense of any of these principles.
- The tourism industry, local authorities and environmental agencies all have a duty to respect the above principles and to work together to achieve their practical realisation.

Source: Department of the Environment (1991: 15)

consciousness and value in its own right, as discussed in Chapter 2. Notably, the recognition that this intrinsic value outweighs its value as a tourism asset may mean that the environment is excluded for tourism use. The first principle also endorses the recommendations of the Brundtland Report by stressing the need for long-term planning considerations as opposed to short-termism.

The principles also stress other elements that are viewed as being essential to achieving sustainable development. The need for a balanced approach to development, respecting the character and nature of the place, is embodied in the principles. This involves the balancing of the needs of the natural environment with the needs of the community, and the needs of the tourist. However, this principle tends to imply that the needs of the tourist as a distinctive group are homogeneous, and that the needs of the local community as a distinctive group are also relatively homogeneous. In reality, this is too simplistic and unlikely to be the case, as there are different types of tourists (as described in Chapter 2) and communities can be divided on grounds of class, ethnicity, religion and gender, all with different preferences for tourism and views on the desirable limits of growth. As Harrison (1996) comments, it is unlikely that the wishes of local people could ever be sufficiently cohesive to offer a practical guide to tourism development.

Nevertheless, community control over development decision-making has often been advocated by those who favour increased local democracy. Ecocentrics support developmental decision-making at local level, not only on democratic principles, but also on the presumption that local people are more likely to act as stewards of the environment than external parties. However, community participation in planning and development may or may not be successful in encouraging people to favour less environmentally damaging development options, as attitudes to the physical environment are likely to reflect economic priorities. Even when tourism is presented as a less environmentally damaging development option than other forms of economic activity, it may not be favoured by the local community. Burns and Holden (1995) comment upon the case of tourism development in the St Lucia Wetlands in Natal, South Africa, an area containing coral reefs, turtle beaches, high-afforested dunes, freshwater swamps, grasslands and estuaries. Rio Tinto Zinc (RTZ), the giant transnational mining corporation, wanted to mine the dunes for titanium dioxide slag. Despite assurances from RTZ over redressive environmental restoration of the area when the mining had ceased, central government was opposed to the use of the area for this purpose on environmental grounds, and instead favoured the development of ecotourism. However, local people, mainly Zulus, favoured the development of mining on the basis that RTZ had a good track record of paying comparatively high wages and investing in schools, clinics and other facilities. The Natal Parks Board, who run the surrounding game parks, were perceived by the local community as paying low wages, and having displaced local people from their lands to establish game reserves in the 1960s and 1970s.

Similarly, at Cairngorm, in the Scottish Highlands of Britain during the 1990s, there was a high level of controversy over the planned development of a funicular railway up the mountainside for the purposes of downhill skiing. Opposition to the scheme from major non-governmental organisations, such as the World Wildlife Fund (WWF) and the Royal Society for the Protection of Birds (RSPB), was based upon the possible environmental impacts on the arctic-alpine environment, which is unique within the British Isles. However, instead of receiving the support of the majority of local people, the two non-governmental organisations were largely seen as outsiders attempting to stop economic development to protect birdlife and flora, thereby denying local people employment and other economic opportunities. To realistically encourage stewardship of the environment by local communities, forms of tourism will need to be developed that are not only sympathetic to the environment, but also offer economic benefits, as is the case in Senegal described in Box 6.4.

A good example of a private sector-led project aimed at encouraging sustainable tourism development is the one developed by the International Federation of Tour Operators (IFTO), in the Balearic Isles, in the Mediterranean. Concerned with the problems that were being caused by tourism development in parts of Mallorca, the largest of the Balearic Isles, the ECOMOST ('Ecological Model of Sustainable Tourism') study was supported by IFTO, the European Community, and the Ministry of Tourism for the Balearics. The aim of the ECOMOST project was to attempt to develop a 'model of sustainable tourism'. The researchers likened this model to tourism's stethoscope, the idea being that it could be used to gauge a destination's attainments against the ideal of sustainability, by using indicators of change. Within the context of the study, the adopted definition of sustainable tourism was 'the maintenance of a balance where tourism runs at a profit but not at the expense of the natural, cultural, or ecological resources' (IFTO, 1994: 6). The report identifies four main requirements for the long-term maintenance of a tourism destination:

Box 6.4

Integrated rural tourism: Senegal

The development of tourism in the Lower Casamance, is an example of how tourism can be used as a tool to enrich the livelihood and well-being of rural peoples. The aim of the scheme was to aid development, and also provide a more meaningful interchange between the local people and tourists than was being experienced on the coast, where hotels had largely been built with foreign capital and local people were excluded from tourist complexes by security guards and high walls.

In total, 13 tourist camps have now been built, from an initial investment of US$7,000 each, provided by the l'Agence de Coopération Culturelle et Technique. Tourists stay in simple lodges, built using traditional materials in the local architectural styles, so diminishing the differentiation between tourist and local facilities. Tourist numbers are restricted to a maximum of 20–40 guests and lodges are only constructed in villages where the population is 1,000 or above. Tourists eat locally grown produce wherever possible, following traditional recipes.

The scheme has proved to be a success, aiding development and social stability, improving health and educational facilities, and critically providing employment opportunities for the young which discourages them from migrating to larger towns to look for employment. Public expenditure of the revenues from tourism is controlled by village co-operatives.

Source: Gningue (1993)

1 The population should remain prosperous and keep its cultural identity.
2 The place should remain attractive to tourists.
3 Nothing should be done to damage the ecology.
4 There should be an effective political framework.

The study developed 'checklists of dangers to sustainable tourism', under four different subheadings: 'population', 'tourism', 'ecology' and 'politics'. For each subsection, different components or targets were identified along with indicators of change, accompanied by 'critical values' which will determine if the indicator is positive or negative, as illustrated in Box 6.5.

Box 6.5

Indicators of sustainable tourism development

Topic	Component or target	Indicator	Critical value
Population	Preserving the population's prosperity	• Population dynamics	• If there is a continuous and major migration of the working population
		• Unemployment rate	• If greater than the national average and/or increasing long term
		• Per capita income	• If lower than the national average
Tourism	Retaining satisfaction of guests and tour operators	Maintenance of quality and monitoring ecology	Persistent and/or significant criticism of destination's condition with special reference to: quality of accommodation, restaurants, service, leisure infrastructure; overcrowding of transport, beaches and sights; ecological condition of nature, the landscape and amounts of waste; aesthetics of townscape, landscape and cultural assets
Ecology	Carrying capacity	• Airport	• If maximum capacity exceeded
		• Tourist attractions	• If roads and parking lots are continuously overcrowded at peak times
		• Drinking water • supplies	• If there is: – water shortage in peak season – long-term danger of salinity, floods, forest fires and other ecological damage
		• Sewage	• If European Union Standards for sewage

Continued

Box 6.5 continued

Topic	Component or target	Indicator	Critical value
			disposal are mostly neglected
		• Protection of species, use of protected areas	• If, due to exploitation by tourists, flora and fauna are becoming imperilled or being destroyed
		• Pollution and emissions	• If, due to tourist use: –water –soil –air –health are continuously being threatened by pollution and/or noise
Politics	Effective tourism and ecologically oriented legislation	Not applicable	Existence of ecologically oriented quality standards

Source: Abridged and adapted from IFTO (1994: 10–15)

The idea of using indicators as an environmental management tool, to help achieve sustainable development, has gained pace since the adoption by the majority of the world governments of Agenda 21 in 1992. For example, the European Union is currently trying to establish relevant information systems, to ensure that socio-economic environments are developed in a way that respects the natural environment. Eurostat, the Statistical Office of the European Communities, in 1997 produced a list of 46 indicators covering areas of the economy, society, environment and institutions. Examples of *economic indicators* that are relevant to the environment include the 'annual energy consumption per capita' and the level of 'environmental protection expenditure as a percentage of GDP'. Examples of *social indicators* include the 'rate of growth of the urban population' and the 'per capita consumption of fossil fuel for motor vehicle transport'. Examples of *environmental indicators* include the 'level of consumption of ozone-depleting substances' and the 'quantity of fertiliser used in agriculture'. An example of an *institutional indicator* is the level of 'expenditure on research and development as a percentage of GDP' (Eurostat, 1997). In 1998, the British government announced a range of indicators to help assess changes in the quality of life of the population, which could be used alongside traditional measures, such as gross domestic product (GDP). Thirteen groups of indicators were decided upon, covering the following areas: economic growth; social investment; employment; health; education and training; housing quality; climate change; air pollution; transport; water quality; wildlife; land use; and waste (Ward, 1998). However, although these developments are encouraging and indicators have undoubted use as part of an environmental management system, both for the general environment and tourism, their usefulness in securing a more sustainable future will ultimately be determined by the political views and commitment of those in power.

Summary

- Sustainable development encapsulates a way of thinking about economic development that emphasises the conservation of natural resources and human development. Development cannot take place upon a deteriorating environmental resource base; neither can the environment be protected when development excludes the cost of its destruction. The main aim of sustainable development is poverty alleviation; that is, satisfying the needs of the world's population but achieving this in a way that does not threaten the earth's resources, or the ability of future generations to satisfy their own needs. Thus it incorporates principles of inter- and intra-generational equity.
- 'Sustainable development' is an ambiguous term and can be interpreted in different ways by groups with opposing political and philosophical viewpoints. Some groups and ideologies believe sustainable development can only be achieved by a radical restructuring of society, involving changes in political structures and value systems. Others believe sustainable development can be achieved through alterations of the market system and improvements in technology that do not threaten the status quo in society.
- The application of the concept of sustainable development to tourism has also led to different perspectives vis-à-vis its meaning. A major

difference is between those who interpret 'sustainable tourism' as advocating the sustaining of tourism in a destination, and those who advocate tourism as a vehicle for achieving sustainable development, which encompasses much wider socially determined goals and priorities. Saarinen (2006) identifies three main traditions of sustainable tourism: 'resource-based', 'activity-based' and 'community-based'.

Further reading

Baker, S., Kousis, M., Richardson, D. and Young, S. (eds) (1997) *The Politics of Sustainable Development: Theory, Policy and Practice within the European Union,* London: Routledge.

Doyle, T. and McEachern, D. (1998) *Environment and Politics,* London: Routledge.

Mowforth, M. and Munt, I. (1998) *Tourism and Sustainability: New Tourism in the Third World,* London: Routledge.

Pepper, D. (1993) *Eco-socialism: From Deep Ecology to Social Justice,* London: Routledge.

Reid, D. (1995) *Sustainable Development: An Introductory Guide,* London: Earthscan.

Saarinen, J. (2006) 'Traditions of Sustainability in Tourism Studies', *Annals of Tourism Research,* 33 (4): 1121-40.

Suggested websites

Earth Institute at Colombia University: www.earthinstitute.columbia.edu/sus_dev

International Institute for Sustainable Development: www.iisd.org

United Nations Division for Sustainable Development: www.un.org/esa/sustdev

United Nations Environment Program: www.uneptie.org/pc/courism/sust-tourism/home.htm

The environmental planning and management of tourism

- Legislation for environmental protection
- Environmental audits and management systems
- Codes of conduct in tourism
- Roles of different stakeholders in the environmental planning and management of tourism

Introduction

The evident need for the environmental planning and management of tourism, in light of the sometimes negative interaction between tourism and the natural environment, has become of concern to governments, NGOs, local communities, donor agencies and the private sector. All of these different stakeholders in tourism have a role to play in influencing the development of tourism and the extent to which its interaction with the environment is positive or negative.

The role of government

As was discussed in the last chapter, emphasis upon natural resource conservation has now become integral to development policy. Through the passing of legislation and use of fiscal controls, governments have potentially a wide range of powers that they can exert upon tourism development with the aim of mitigating negative environmental impacts. However, the prioritisation given by a government to environmental protection and conservation will be a reflection of its own philosophy regarding the values of nature. The typical government prioritisation of nature conservation in a hierarchy of national goals is shown in Figure 7.1.

O'Riordan's (1981) hierarchy suggests that concern for the natural environment is only likely to happen after national security and economic

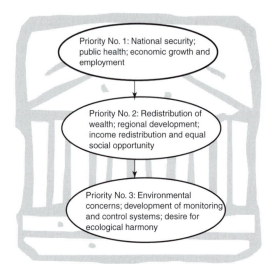

Priority No. 1: National security; public health; economic growth and employment

Priority No. 2: Redistribution of wealth; regional development; income redistribution and equal social opportunity

Priority No. 3: Environmental concerns; development of monitoring and control systems; desire for ecological harmony

Figure7.1 *A hierarchy of national goals*

development have been achieved. This is reflected in tourism development policy, which has traditionally emphasised economic objectives, such as employment creation and regional development, whilst treating the environment in an instrumental fashion.

Although there now exists a higher level of awareness amongst policy-makers over the need to pay more attention to natural resource management and conservation than previously, there still exists conflict over the extent to which stewardship of the environment should detract from economic growth. An allegation made by many developing countries against the western world, which seemingly wishes to dictate environmental policy on a global level, is that the West has the luxury to be able to be concerned over the environment because it has largely fulfilled many of its national goals. As Stone (1993: 42) comments: 'Joan Martin Brown, an official with UNEP, has remarked: "The Third World says you're telling us not to do what you did to achieve your high standard of living. What are you going to do for us?"' It is therefore likely that the extent to which governments are willing to pursue policies where the conservation of the natural environment is made a priority will depend greatly upon the ability of sustainable development to meet economic objectives. This will be reliant upon the dissemination of knowledge based upon models of good practice, advancement in environmentally benign technologies, and a revised market system that incorporates the costs of negative environmental externalities in the prices of goods and services.

Pressure to use natural resources in an unrestricted fashion for wealth creation is exacerbated within the global economic environment. By encouraging the deregulation of trade and the liberalisation of economies, it is difficult for governments to refuse foreign investments, particularly in situations where countries are faced with a large variety of social and economic problems. Instrumental to the encouragement of this process have been international trade agreements, notably the GATT (General Agreement on Tariffs and Trade) and GATS (General Agreement on Trade in Services), which are especially relevant to tourism. These agreements are administered by the World Trade Organization, which has been operational since 1995. A basic premise of GATS is free-market access, and the denial of any protectionist measures, in effect opening up countries to unlimited amounts of foreign investments. An example of how a deregulated global economy can impact upon natural resources is given in Box 7.1, concerning the granting of logging concession rights in the Democratic Republic of Congo.

Box 7.1

Logging in the Democratic Republic of Congo (DRC)

The village of Lamoko in the Democratic Republic of Congo is adjacent to an extensive stretch of virgin rainforest covering thousands of hectares. In 2005, representatives of a major timber firm arrived to negotiate an agreement with the traditional landowners. The chief, who received no legal advice and did not realise that one tree could be worth approximately US$8,000 in North America or Europe, signed away his community's rights to the forest resources for 25 years.

The company gained rights to log tens of thousands of exotic hardwoods, e.g. *Afromosia* (African Teak) in return for promising to build Lamoko and other communities in the area three village schools and pharmacies. They also promised the chief 20 sacks of sugar, 200 bags of salt, some machetes and a few hoes. Although logging roads have been built deep into the forests and the company has started chopping and exporting trees, no schools or pharmacies have yet been constructed. One man commented that they asked the company to provide wood for the villagers' coffins but it refused.

The village of Lamoko is not an isolated case. Other villages have also given up their rights to the forests in return for sugar, salt and tools. Without the trees and the forests, upon which the communities have traditionally based their livelihoods, they face poverty. Although the companies are obliged to employ local people, they usually bring in their own people, leaving local people with unskilled jobs paying less than US$1 per day. Another concern of unsustainable logging of tropical rainforests is that they provide important carbon reserves. It is now thought that the burning of the tropical rainforest is a major source of greenhouse gas (GHG) emissions.

Source: Vidal (2007)

The ability of transnational investors to transfer capital between countries means that many governments are reluctant to place any obstacles, such as environmental regulations, in the way of the inflow of capital for tourism development projects. For example, the demand of government X for mandatory environmental impact assessments on large-scale tourism development projects may mean costly delays to a foreign investor, inducing a longer payback period and a reduced rate of return on their investment. Consequently, the investor may decide to place their investment in country Y, where environmental regulations are more lax or non-existent. Country X now loses the potential economic benefits from the project, and the government will also undoubtedly court political unpopularity for refusing the investment. The environment may consequently come to be viewed by some of the populace as a hindrance to development, rather than something to be cherished and forming the basis of their long-term prosperity.

Additionally, debt repayments to western banks and governments, combined in some cases with strict external controls on government expenditure associated with 'structural adjustment' of some national economies demanded by the International Monetary Fund (IMF), greatly influence the ability of countries to allocate resources for environmental protection. 'Structural adjustment' refers to a set of free-market economic policies imposed by the World Bank and IMF as a condition of receiving financial help. Part of this adjustment involves a lessening of government spending to reduce a country's debt level, the ultimate aim being to minimise state economic intervention and to generate market-orientated growth. Consequently, if conservation of the environment is low on the list of government priorities, it may be one of the first areas to be cut back in government expenditure.

Certainly where a government favours no kind of regulation of commercial activity, then the pace of development is likely to be quicker, but the consequences for the environment will be less easy to monitor and are more likely to be destructive. One example of the potential consequences of inward investment for tourism with little government control is the case study of Langkawi in Malaysia, as described in Box 7.2.

Although national priorities and global economic forces may detract from environmental conservation, governments do have a range of policy and legislative measures at their disposal which can be used to safeguard environmental resources, if they wish to use them. Typical measures that

---

Box 7.2

Uncontrolled inward investment for tourism development in Malaysia

In the 1980s, the Malaysian government decided that Malaysia needed to have a well-known destination to help promote its tourism. Subsequently, it decided to develop one of the archipelago of islands known as Langkawi, situated off the west coast of peninsular Malaysia. A decision was taken to develop a tourism complex to be named the 'Langkawi Resort' on 1,417 hectares of land around a beautiful bay named Tanjung Rhu, at a total cost of US$1 billion in 1980s prices. This bay was specifically chosen because of its outstanding beauty, including casuarina trees, a beach front carpeted with white flowers, and its crystal-clear lagoon and waterways.

The development company was called 'Promet', the largest shareholder being Singaporean, and financial support was also given by the federal and state governments for infrastructure development, including an international airport. Initial construction work began in 1984, but by 1985, Promet was put into receivership. Today Tanjung Rhu is a devastated area resembling a wasteland or moonscape, with jungle and mangrove swamps having been completely cleared, the sand having been removed from the beach for use in construction, and the water being silty and unclear.

Source: Bird (1989)

---

governments at national, regional and local levels can consider include a mix of policy and planning measures, such as:

- The establishment of protected areas through legislation, for example, national parks, and application for international recognition of significant environments, such as the World Heritage Site (WHS) status given by UNESCO
- The implementation of land-use planning measures such as zoning, carrying capacity analysis, and limits of acceptable change (LAC) to control development
- Mandatory use of environmental impact analysis (EIA) for certain types of projects
- Encouraging co-ordination between government departments over the implementation of environmental policy, and entering into dialogue with the private sector to encourage the adoption of environmental

management policies, such as environmental auditing and the development of environmental management systems.

Protected areas

Governments possess the powers of legislation to establish protected areas and a variety of different types of designations exist. Based upon work carried out under the patronage of the United Nations Environmental Program (UNEP), the World Tourism Organisation (1992) has produced several classifications of protected areas. Arranged by order of the degree of their permitted use by humans, the most restrictive being at the top and the less restrictive at the bottom, they are as follows:

- *Scientific Reserve/Strict Nature Reserve* – to maintain and protect the existing ecological balance of the area for scientific study and for environmental education
- *National Parks* – to protect outstanding natural and scenic areas for educational, scientific and recreational use. Generally they tend to cover large areas of land, are not materially altered by human activity, and extractive industries are not permitted inside their boundaries
- *Natural Monuments/ Natural Landmarks* –to protect and preserve nationally significant natural features defined by their special interest or unique characteristics
- *Managed Nature Reserve/Wildlife Sanctuary* – human manipulation, for example, the culling of a predatory species, is involved in these areas, to ensure the protection of nationally significant species of the biotic communities and physical features of the landscape
- *Protected Landscapes* – to maintain nationally significant landscapes which are characteristic of the harmonious interaction of humans and nature. Emphasis is also placed upon the enjoyment of the area through recreation and tourism, as long as this does not detract from the normal lifestyle and economic activity of the area
- *Resource Reserve* – to protect or sustain the resources of an area for future use by prohibiting development activities that might threaten them
- *Naturally Biotic Area/Anthropological Reserve* – to permit the way of life for societies that are living in harmony with the environment to continue, uninterrupted by modern technology and human activity
- *Multiple-use Management Area/Managed Resources* – for the sustained production of a mix of water, timber, wildlife, pasture and outdoor

recreation. The conservation of nature is orientated to the support of economic activities, although specific zones can also be designed within these areas to achieve specific conservation objectives.

For some of these protected areas, tourism has direct relevance in terms of both benefiting from the protection of the environment from other forms of development and, if planned and managed appropriately, being able to make a positive economic contribution to environmental protection. For instance, carefully regulated and managed tourism, such as small groups interested in scientific education, could help fund the research and protection of a scientific reserve, whilst tourism revenues have already directly aided the establishment of national parks, especially in developing countries.

One of the most common protected area designations worldwide, and one in which tourism plays an important role, is the national park. The responsibility for passing legislation to establish a national park lies with governments. They are usually established with the objectives of protecting outstanding natural areas from overdevelopment and providing areas of access to nature for tourists and recreationists. The first national park in the world was created at Yosemite in America in 1872, and it is no accident that its establishment coincided with the increasing urbanisation of America, in the latter half of the nineteenth century. The focus of the rationale for establishing the Yosemite National Park was based not solely upon conservation but upon the provision of nature for the enjoyment of urban dwellers.

THINK POINT

What are the benefit of establishing national parks? Are they principally to protect the environment or to encourage tourism?

National parks may also play a greater part in the national psyche than being simply areas for nature conservation and recreation. Hall and Lew (1998) suggest that the development of the Yosemite National Park, and the need for a rapidly urbanising American population to stay in contact with nature, was to act as a reminder of the pioneer mentality that modern America had been built around, as much as for recreational purposes. Instrumental to encouraging urbanised Americans to visit the mountains was John Muir, the founder of the Sierra Club in the USA, who thought that visiting mountains was good for one's soul (Eagles et al., 2002). An interesting alternative perspective on why national parks were created is also given by MacCannell (1992: 115):

> The great parks, even great urban parks, Golden Gate in San Francisco or Central in New York, but especially the National Parks, are symptomatic of guilt which accompanies the impulse to destroy nature. We destroy on an unprecedented scale, then in response to our wrongs, we create parks which re-stage the nature/society opposition now entirely framed by society.

In MacCannell's view, the creation of national parks is symptomatic of a release of collective guilt over the harm humans have caused to the non-human environment, during the process of industrial modernisation. The theme of guilt is also inherent to the American poet Catlin's advocation in the 1830s of the need to establish a nation's park in response to the destruction of aboriginal cultures as a consequence of development (Eagles *et al.*, 2002). By the end of the nineteenth century, national parks had been established in other countries, including Banff National Park in Canada in 1887 and Tongariro National Park in New Zealand in 1894. Common features of the emerging national parks at the end of the nineteenth century were that they were created by government action; they consisted of large areas of natural environments; and they were available to all people (ibid.).

However, MacCannell (1992) suggests that although the creation of the parks represents a 'good deed' of industrial civilisation, their creation affirms the power of humans over nature. The collective conscious or subconscious guilt held by western society over the way it has instrumentally used nature in the process of development may also influence perceptions of tourism. The fact that tourism often takes place in relatively unspoilt environments helps us to imagine how nature and life may have been before industrialisation and development took place. It may therefore be difficult to perceive that tourism can be environmentally threatening, as many of the environments where it takes place represent the antithesis to what is deemed to have gone wrong in developed societies.

Ironically, the biggest threat today to some national parks in the developed world is tourism. This was recognised by Murphy over 20 years ago (1985: 41):

> In addition to serving the drive-through visitor, Yosemite accommodates those who wish to stop-over, providing campsites, cabins, and a hotel in Yosemite Village. At times the valley which contains this village and attracts most of the visitors becomes an

off-shoot of Los Angeles, complete with traffic jams and smog conditions – the very conditions tourists are trying to escape.

Over-popularity is perhaps one of the greatest hidden dangers to the natural environment that the parks are trying to protect. Indeed, by giving an area the status of a national park, it automatically becomes an attractive place to visit. This can be especially problematic for park management in areas where national park status was given several decades ago, but where the pace of technological change and transport development has meant that they have become easily accessible to large numbers of people for day trips and vacations.

The creation of national parks is not restricted purely to terrestrial areas of the world's surface. For instance, marine parks have also been established to protect coral reefs in places such as the Netherlands Antilles (Bonaire Marine Park) and in the Seychelles (Saint Anne National Marine Parks). Ultimately, the type of national park or protected area that is established in a country is likely to be a reflection of a variety of different factors, including existing levels of economic development, population density, and the extent of institutional and financial support from government.

In developing countries, the rationale for the establishment of national parks has been much more closely associated with the conservation of wildlife, supported by the revenues from tourism in some cases, rather than providing recreational opportunities for urban dwellers (see chapter 4). National parks act as important focal points for attracting international tourists, as in east and south Africa, Costa Rica, India, Nepal and Indonesia (WTO, 1992). Growth in domestic tourism numbers to national parks is also significant in many developing countries. However, sometimes the creation of national parks has adverse cultural impacts, including the displacement of indigenous peoples, as was discussed in Chapter 3. Although national parks are very important for aiding conservation of the natural environment, their creation can result in costs besides benefits, as summarised in Box 7.3. The key to achieving success in national parks is the development and implementation of suitable management plans, which balance the use of natural resources with the needs of the local people, and the expectations of the tourists.

Besides legislating for the creation of national parks, other types of protected area status may be used by governments, as exemplified in the case of the Balearic Isles in Box 7.4.

Box 7.3

Summary of the costs and benefits of national parks

Benefit	Cost
Protects landscapes, wildlife and ecological communities	Unless carefully managed, recreation and tourism can pose a threat to both the landscape and wildlife that the park was established to protect
Provides a place for people to have access to and experience nature. Tourists can also provide revenues for park management, scientific research and conservation projects	Granting of national park status focuses attention on the area. This may lead to the attraction of too many tourists and overcrowding of the area
Offers employment opportunities for local people to become involved in conservation of the environment rather than destructive practices such as clearing natural vegetation for agriculture and poaching	Indigenous peoples can be excluded from their territory to protect landscape and wildlife

A further form of protected area status that is having increasing significance in the context of tourism are World Heritage Sites (WHSs), a status confirmed by the United Nations Environment and Scientific Committee and Organization (UNESCO).

THINK POINT

How have the processes of the Industrial Revolution and urbanisation influenced the creation of the national parks?

World Heritage Sites

The catalyst to the concept of World Heritage Sites (WHSs) was the construction of the Aswan High Dam in Egypt and the subsequent threat

Box 7.4

The use of legislation for protected area status in the Balearic Isles

The growth of tourism in Spain since the unveiling of Franco's 'National Stabilization Plan' of 1959 established a policy of *'crecimiento al cualquier precio'* (growth at any price). The pressure this policy placed on the environment in the Balearic Isles of Spain led the Balearic Isles' government to pass a series of laws in 1991, to limit new tourism construction in both the physical and urban environments. Under the new legislation, one-third of Mallorca's surface area is now protected from future development, based upon protective legislation of three different types:

- *Natural Areas of Special Interest* – areas deemed to be of outstanding natural value and ecological importance
- *Rural Areas of Scenic Interest* – includes areas of primarily traditional land-use activity, but that are still deemed to be of special scenic value.
- *Areas of Settlement in a Landscape of Interest* – includes areas of a primarily urban nature although declared to be of exceptional scenic value.

In all these areas, no new construction will be permitted, beyond that of the existing land-use and infrastructure requirements, for example, urban regeneration projects.

Source: Gamero (1992)

to the Abu Simbel temples from the flooding of the valley in 1959. After an appeal from the governments of Egypt and Sudan, UNESCO launched an international safeguarding appeal, with over half the campaign costs of US$80 million coming from other countries (UNESCO, 2006a). The necessity to help protect sites containing significant cultural and natural heritage led UNESCO to establish the Convention for World Heritage Sites in 1972. Central to the understanding of heritage is the linking together of natural and cultural environments, as specified by UNESCO (2006b).

Today there are 830 designated WHSs, of which 644 are classified as being cultural, 162 as natural and 24 as mixed (UNESCO, 2006b). They are spatially diverse, being found in 138 nations, and vary from sites that are fairly remote from human habitation to ones that include living communities. However, the World Heritage Fund has less than US$4 million per annum to support these sites, meaning that alternative sources of income have to be found to support their management. One possible means of raising sources of income is tourism.

One outcome of the designation of a site on the World Heritage List, similar to the creation of a national park, is that it publicises it, emphasising that there is something worth travelling to. It consequently may become a focus for tourism, offering economic opportunities but also leading to problems of tourism management. The issue of balancing economic opportunity with the need for conservation has special relevance to iconic sites such as the Parthenon in Athens; Angkor in Cambodia; the Taj Mahal in India and Machu Pichu in Peru. There is a danger of such sites becoming items for tourists to tick off ('been there ... done that'), and a subsequent structuring of the operations of the site to cater for mass tourism, presenting a threat to its integrity and conservation. There is also a threat to sites that are at a much earlier phase of development, as the granting of WHS status may be significant in pushing a new site along the lifecycle of tourism development. A real danger is that such sites may be less equipped to respond to tourism than more iconic sites, many of which have a long experience in dealing with visitors, even before their designation as WHS.

The economic benefits that WHS designation can bring to the area of a natural or cultural site are those generic to tourism. A variety of economic multiplier effects, including income and employment, will be experienced as a consequence of tourism visitation. Other types of economic benefits include a macro-level contribution to the balance of payments and to gross domestic product. For some communities, notably in the rural areas of developing countries, where development opportunities may be limited, tourism may be the most viable option for material advancement.

Pressures for overuse may exist, as there may be a temptation for countries to market WHS sites as iconic attractions for the reason of economic benefits. They may also arise because of the availability of increased information about a site and the easing of accessibility to it, through improved infrastructure. The influence of information and accessibility as factors of tourism demand is as relevant to WHSs as it is to other types of tourism destinations. The propensity to visit these sites has been increased by the advent of cheap air travel and information technology, therefore placing even greater pressure on WHSs that are located in close proximity to major tourism-generating zones. For example, Pompeii in Italy received 863,000 visitors in 1981, rising to 2 million by 2006. Inevitably the advent of mass tourism to WHSs, characterised by the involvement of tour operators and organised groups, leads to visitor management pressures.

Tourism that is not carefully planned may pose a threat not only to the physical and cultural entity of the WHSs but also their continued designation. Along with problems of climate change, urbanisation and war, tourism is a potential problem. For example, in a report by the Centre for Future Studies (CFS), it is suggested that the WHSs of Australia's Great Barrier Reef and Kathmandu could be unpopular by 2020, owing not only to climate change and urban development respectively, but also to the presence of too many tourists. The awareness of the problems that tourism may create led the World Heritage Committee to launch in 2001 the World Heritage Sustainable Tourism Program. The programme covers different aspects as listed:

1 Building the capacity of site management in dealing with tourism, notably through the development of a Sustainable Tourism Management Plan
2 Training local populations in tourism-related activities so that they can participate in and receive benefits from tourism
3 Helping to promote relevant local products at the local, national and international levels
4 Raising public awareness and building public pride in the local communities through conservation outreach campaigns
5 Attempting to use tourism-generated funds to supplement conservation and protection costs at the sites
6 Sharing expertise and lessons learned with other sites and protected areas
7 Building an increased understanding of the need to protect World Heritage, its values and policies within the tourism industry.

(UNESCO, 2006b:1)

Although there is an inherent emphasis on the relationship between economic benefits, conservation and community involvement, as would be expected in any strategy or plan purporting to sustainable principles, the problems of community support for a WHS may extend beyond purely its involvement or economic benefit from it. It can often be a question of how their heritage is being presented to tourists and who controls it. Alongside the problematic issue of how to define and identify a 'community' (as referred to in the last chapter), issues of politics, power, representation and authenticity mean that how culture is presented is sometimes contentious. It is not necessarily the case that WHS designation will automatically have community support.

For example, community dissension to the designation of World Heritage Status is a theme that is relevant in the context of the Wadden Sea in the Netherlands. Exploring local community reaction to the suggestion of the designation of the area as a WHS, there was found to be widespread opposition to the proposal (Bart *et al.*, 2005). The main rationale for opposition was based upon the perceived loss of control in local decision-making by the community as powers were transferred to UNESCO. Closely linked to this was a fear that the local area would be given global significance by this designation, leading to increased pressures with little financial support to aid its management.

Land-use planning methods

The pressures that can be placed on destinations and protected areas from tourism makes its planning and management of the utmost importance, both for the conservation of natural and cultural resources, and for the securing of the benefits of tourism into the future. The next section of this chapter considers the range of planning and management techniques that are available to control any negative consequences of tourism upon the natural environment.

Zoning

Zoning is a land management strategy that can be applied on different spatial scales, for instance, within a protected area, or at a regional and even national level. According to Williams (1998: 111):

> Spatial zoning is an established land management strategy that aims to integrate tourism into environments by defining areas of land that have differing suitabilities or capacities for tourism. Hence zoning of land may be used to exclude tourists from primary conservation areas; to focus environmentally abrasive activities into locations that have

been specially prepared for such events; or to focus general visitors into a limited number of locations where their needs may be met and their impacts controlled and managed.

Thus, zoning can provide a proper recognition of the resources that exist in the area and subsequently identify where tourism can and cannot take place. With specific reference to the use of zoning in protected areas, the World Tourism Organizaton (1992: 26) remarks:

> a protected area can be divided into zones of strict protection (a 'sanctuary zone', where people are excluded), wilderness (where visitors are permitted only on foot), tourism (where visitors are encouraged in various compatible ways), and development where facilities are concentrated.

Typically, zoning an area involves two key stages. The first or 'descriptive' step involves identifying important values and recreational opportunities, subsequently necessitating or leading to the production of an inventory of resource characteristics and types of recreational opportunities. The second or 'allocation' step involves deciding what values and opportunities should be made where in protected areas (Eagles *et al.*, 2002).

An example of how zoning has been used in an attempt to balance the requirements of scientific research, conservation, tourism and other forms of commercial activity is that of the Great Barrier Reef Marine Park in Australia, as described in Box 7.5.

Another example of land-use zoning takes place in the Canadian National Park system. The Canadian Parks Service manages 34 national parks, and one marine park, covering a total area of 180,000 square kilometres. Five zones have been designated for application in the national parks, categorised by the resource base of the area and the amount of recreational access that is allowed there, as follows:

- *Zone 1: 'special preservation'* – areas that contain strictly protected rare or endangered species and where access is strictly controlled
- *Zone 2: 'wilderness'* – represents 60 to 90 per cent of the park area and the primary aim is resource preservation. Use is dispersed with only limited facilities
- *Zone 3: 'natural environment'* – this area acts as a buffer zone between zones 2 and 4 and access is primarily non-motorised

Box 7.5

Zoning in the Great Barrier Reef Marine Park, Australia

The Great Barrier Reef in Australia forms the world's longest coral reef, stretching for almost 2,000 kilometres along the north-eastern coast of Queensland. The reef is home to approximately 350 species of coral, 1,500 types of fish and six species of turtle. The development of international airports at the towns of Cairns and Townsville, which are conveniently situated for access to the reef, has meant that the number of tourists wanting to visit the reef has grown substantially since the late 1970s. The growth of tourism and other economic activities based upon the reef has meant that increased pressure is being placed upon it, and also that there is increased potential for conflicts between different user groups such as fishermen, tour operators (who can take groups of several hundred tourists on large catamarans and other boats to the reef), and recreationists such as scuba divers. The types of consequences that result from the use of the reef for tourism include: physical damage from anchors, moorings, snorkelling, diving and people walking on it; collecting of marine fauna; and the discharge of waste, litter and fuel.

The response to these problems was the establishment of the Great Barrier Reef Marine Park Authority (GBRMPA) to co-ordinate the management and development of the area. One of their functions was to zone the park to allow multi-use of the reef whilst preserving its ecology. They developed four different types of zones:

1 *Preservation Zones* – areas in which use of the reef for virtually any purpose is prohibited

2 *Scientific Research Zones* – areas where scientific research is permitted under strict control

3 *Marine National Park Zones* – areas where scientific, educational and recreational uses are permitted

4 *General Use Zones* – areas where some commercial and recreational fishing is permitted.

Commercial tourism is permitted in zones 3 and 4, and the zoning process also includes the designation of Special Management Areas, in which reefs that are being intensively used for tourism or other purposes can be protected or conserved.

Another aspect of the GBRMPA's role is the environmental impact management of the reef. All proposed tourist operations are subjected to environmental assessment before they can be granted a permit to operate on the reef, and large-scale developments, or those that are assumed to produce unacceptable environmental impacts, have to prepare environmental impact statements.

Source: Simmons and Harris (1995)

- *Zone 4: 'recreation'* – overnight facilities such as campsites are concentrated in this area
- *Zone 5: 'park services'* – this area is highly modified, providing many services but represents less than 1 per cent of the park area.

Areas like zone 5,which are highly modified to suit the needs of certain types of tourists or recreationists, often encompassing a range of visitor attractions to persuade people to remain there instead of venturing into more environmentally vulnerable areas, are often referred to by recreational planners as 'honeypots'. Besides being used within protected areas, honeypots can also be developed on a wider spatial scale, to stop tourism developing in areas of regions or countries where the nature has been identified as being significant.

The benefits of zoning in terms of the conservation of natural resources can be summarised as: (i) defining the types of tourism and other recreational activities suitable to different spatial areas and ecosystems; (ii) aiding the management of visitor impacts to ensure they do not exceed acceptable levels; and (iii) increasing the awareness of tourism stakeholders of the environmental values of the ecosystem and natural resources (after Eagles *et al.*, 2002).

Carrying capacity analysis

One of the techniques that has been most commonly referred to in the literature on tourism planning is that of 'carrying capacity analysis'. The origins of this concept can be traced back to the last century, when concerns were being expressed over the levels of wildlife population that could be supported by the environment. In more recent times, the concept has been applied to tourism. The World Tourism Organization (1992: 23) defines carrying capacity as:

> fundamental to environmental protection and sustainable development. It refers to maximum use of any site without causing negative effects on the resources, reducing visitor satisfaction, or exerting adverse impact upon the society, economy and culture of the area. Carrying capacity limits can sometimes be difficult to quantify, but they are essential to planning for tourism and recreation.

Similarly, Mathieson and Wall (1982: 21) state: 'Carrying capacity is the maximum number of people who can use a site without an unacceptable alteration in the physical environment and without an unacceptable decline in the quality of the experience gained by the visitors.'

From these definitions it is evident that there are different elements to the concept of carrying capacity beyond physical considerations. According to Farrell (1992), there are at least four different types of carrying capacity, and O'Reilly (1986) identified economic, psychological, environmental and social carrying capacities as being of relevance to tourism. All have threshold levels beyond which the carrying capacity would be deemed to have been exceeded, leading to deterioration in the quality of the aspect under consideration.

The four types of capacity are:

- *Economic carrying capacity* – the extent of the dependency of the economy upon tourism
- *Psychological carrying capacity* – the expressed level of visitor satisfaction associated with the destination
- *Environmental carrying capacity* – the extent and degree of impacts of tourism upon the physical environment
- *Social carrying capacity* – the reaction of the local community to tourism.

The four carrying capacities are not independent of each other, but it may be possible to exceed the threshold limit of one capacity for a limited amount of time, without there being necessarily a detrimental effect upon another type of capacity. For example, it is possible that an increase in the number of trekkers in a mountain area could lead to increased levels of destruction of flora from trampling, even threatening the ecological balance of the area, whilst the satisfaction of the visitors is not diminished. However, if the number of walkers continued to increase and damage to the environment increased proportionately, eventually the level of environmental damage would lead to a threshold level being crossed, whereby it detracted from the level of satisfaction with the wilderness being experienced by the walkers. The whole notion of when damage occurs is debatable, as Wight (1998: 78) remarks:

> The term damage refers to a change (an objective impact) and a value judgement that the impact exceeds some standard. It is best to keep these two separate. In terms of human impact, a certain number of hikers may lead to a certain amount of soil compaction. This is a change in the environment, but whether it is damage depends on management objectives, expert judgement and broader public values.

Early attempts in the field of tourism planning at identifying the carrying capacity of destination areas were preoccupied with trying to quantitatively determine the number of tourists that could be accommodated in an area, without causing 'unacceptable' environmental and social changes. Although the concept of carrying capacity had been evolving and developing in the field of recreation studies since the 1960s, it has only been since the late 1980s that the concept has become of interest to tourism researchers and planners. One notable exception to this observation was a study carried out for the Irish Tourist Board by the United Nations in 1966, which attempted to define the numbers of visitors that different destinations in Donegal in Ireland could tolerate, without harming the physical environment (Butler, 1997). However, owing to the highly complex nature of tourism, the notion of quantifying capacity limits is extremely problematical, partly because it will be influenced by a variety of factors, as listed in Box 7.6.

Setting capacity limits across environmental, economic and social areas will inevitably involve value judgements. For example, it may be viewed

---

Box 7.6

Factors that will influence the carrying capacities of tourism destinations

- Fragility of the landscape to development and change
- Existing level of tourism development and supporting infrastructure, for example, sewage treatment facilities
- The number of visitors
- The types of tourists and their behaviour
- The degree of emphasis placed on the environmental education of tourists and local people
- Economic divergence and dependency upon tourism
- Levels of unemployment and poverty
- Attitudes of local people to the environment and their willingness to exploit it for short-term gain
- The existing level of exposure of cultures and communities to outside influences and other lifestyles
- The level of organisation of destination management.

as acceptable by decision-makers to exceed cultural and environmental carrying capacity limits to maximise economic benefits, although the long-term sustainability of such a decision is questionable. Conversely, as is the situation in the kingdom of Bhutan, where only a few thousand tourists per annum are permitted in order to protect the physical and cultural environments of the country, the opportunity to maximise the economic benefits from tourism is missed.

Political reasons may also make it difficult for fixed carrying capacities to be operationalised, as such an initiative would be reliant upon government intervention. Butler (1997) suggests that it would be politically unacceptable to the private sector for governments to intervene and regulate the capacity of destinations, as tourism represents a form of free enterprise of capitalism and competition, and the private sector is generally opposed to external control by government. Additionally, given the disparate nature of the resources used by tourists in destinations, and in some cases an absence or lack of clear ownership of these resources, the responsibility for their management is highly problematic. It is this conceptual fragmentation that makes it extremely difficult to operationalise.

Today the notion that there is a fixed ceiling, a threshold number of visitors which tourism development should not exceed, has been largely discredited (WTO, 1992; Williams and Gill, 1994; Saarinen, 2006). Coccossis and Parpairis (1996: 160) comment:

> However, until our understanding of the interactions between the environment and development – human actions – is much more profound, the concept of carrying capacity cannot be used in planning and practice as an absolute tool offering exact measurements but, instead, as one that is under continuous revision, development and research.

Owing to the difficulty of quantification and fixed carrying capacity limits, more emphasis is being placed on indicator monitoring systems to identify potential problems, rather than trying to set absolute numerical limits of tourist numbers for destinations.

Limits of acceptable change

An evolution of the technique of carrying capacity is the 'limits of acceptable change' (LAC) or alternatively called the 'limits of acceptable use'. According to McCool (1996: 1):

> The Limits of Acceptable Change (LAC) planning system was developed in response to a growing recognition in the US that attempts to define and implement recreational carrying capacities for national park and wilderness protected areas were both excessively reductionist and failing.

As is indicated in the above definition, the LAC system, like carrying capacity, has its roots in wildlife management and recreational planning. It is only comparatively recently that the technique has begun to be talked about within the context of tourism planning, and its application to the field is at present limited. A deficiency of carrying capacity analysis, as pointed out in the preceding section, is that many of the problems associated with tourism are not necessarily a function of numbers but rather of people's behaviour. The advantage of the LAC system is that it does not attempt to quantify the numbers of tourists that can be accommodated in the area. Instead the premise of the LAC system is the specification of the acceptable environmental conditions of the area, incorporating social, economic and environmental dimensions, and also its potential for tourism (Wight, 1998). The system is therefore reliant upon identifying the desired social and environmental conditions in an area.

The mechanics of the LAC system involve the adoption of a set of indicators which are reflective of an area's environmental conditions, and against which standards and rates of change can be assessed. Typically, the indicators would relate to the state of the destination's natural resources, economic criteria, and the experiences of local people and tourists. The indicators would therefore be a mix of scientific and social measures. For example: the levels of water, air and noise pollution could be monitored; the percentage of the workforce employed in the tourism sector assessed; crime rates and driving accidents associated with tourism recorded; and levels of tourist satisfaction evaluated. Such indicators would be symptomatic of the impact tourism is having within the destination, and the effect it is having on the quality of life of residents. The indicators should be regularly monitored and evaluated, and strategies identified by the managing authorities to rectify any problems, and to progress towards the desired environmental and social conditions that the LAC system is intended to help achieve. It is important to point out that, owing to the nature of the indicators, measurement cannot be purely scientific, but is also dependent upon a citizen input besides a professional one. As the name suggests, LAC accepts that some change is inevitable, and provides a framework to monitor that change.

Environmental impact analysis

Box 7.5 briefly referred to the role of environmental impact assessments (EIAs) and environmental impact statements (EISs) within the context of the environmental management of the Great Barrier Reef in Australia. Environmental impact assessment, as is suggested in its name, is concerned with assessing the predicted effects of development upon the environment and thereby providing decision-makers with information on the likely consequences of their decision to proceed with a development. The origins of EIAs as a formal part of planning procedure can be traced to the passing of the National Environmental Policy Act (NEPA) in the USA in 1969, which required the preparation of EISs by federal agencies for all major projects. It is not a coincidence that the adoption of NEPA was also accompanied by increasing vocality from environmental groups over the effects of development upon the environment. Weston (1997: 5) comments: 'Indeed the introduction of EIA through the American NEPA was as much a response to political pressure from the growing environment lobby as it was an attempt to introduce a new planning technique.'

Since the passing of the NEPA, EIAs have become a widely discussed and used planning instrument to try to assess the likely consequences of development. The use of environmental assessment can vary from site-specific development to forming part of a strategic environmental assessment (SEA) aimed at examining the consequences of environmental policy. There is no set structure to the components of an EIA but it is generally assumed that EIAs would typically assess future levels of noise pollution, visual impact, air quality, hydrological impact, land-use and landscape changes associated with a development. Most EIAs involve five stages: identification of the impact; its measurement; interpretation of the significance of the impact; displaying the results of assessment; and the development of appropriate monitoring schemes. Importantly, Weston draws attention to the point that within the EIA process, there needs to be monitoring of the impacts during the construction/implementation and operational stages, and also auditing of the EIA process by comparing predicted and actual impacts. The different types of techniques for carrying out an EIA include the use of matrices, overlay analysis, adaptive environmental assessment and management (AEAM), and systems and network diagrams.

The types of tourism developments that would be subject to environmental impact assessments could include hotel complexes, visitor attractions, marinas, and associated infrastructure such as airports, roads, waste treatment and energy plants. The growing importance of tourism as a force of environmental change is recognised in the inclusion of tourism complexes in legislative requirements of the types of projects to be subjected to environmental impact assessment. For instance, in the Republic of Korea since 1981, tourism complexes have been included along with water and energy plants, industrial areas, ports, railways, roads and airports as requiring environmental impact assessments. Similarly, all planned hotel or resort facilities with more than 80 rooms adjacent to environmentally sensitive areas such as rivers, coastal areas, lakes and beaches, or in the vicinity of national parks, are subject to EIAs in Thailand (Wathern, 1988). For countries in the European Community, since the adoption in 1988 of the European Commission Directive (85/337/EEC) formulated in 1985 and later updated in the 1997 EU Directive (97/11/EC), it is recommended that all ski-lifts, cable cars, roads, harbours, airfields, yacht marinas, holiday villages and hotel complexes be subjected to environmental impact assessment.

However, although EIAs may seem to be a good idea, they do face a number of problems in their implementation. One major problem is the cost of preparing an environmental impact statement, which will require a variety of specialists including geologists, hydrologists, geographers, environmental scientists, and possibly sociologists and anthropologists if the analysis is to incorporate a social impact statement (SIA). If the costs are to be borne by the developers, then it may put off some developers from going ahead with the scheme. If the EIA is paid for by the developer, there will also be the question of its ownership and objectivity, which may bring its findings into disrepute. A further problem is predicting the timing of the impacts and when they are going to occur. Therefore it is necessary to distinguish between the impacts that will occur during construction, operation, and possible closure of the complex. Additionally, within the context of tourism development, there is the additional problem that the majority of developments consists of small-scale enterprises which are not subjected to environmental impact assessment, and where the subsequent nature of environmental damage is incremental and cumulative (Butler, 1993).

One method that attempts to deal with the problem of cumulative impacts is the cumulative effects assessment (CEA). This can refer to either the on-going effects of one particular project or the combined effects of a range of different projects. Certainly this technique would seem to have a direct relevance to tourism. However, it is as yet limited in its methodological approach, and there is a lack of information in established texts on environmental impact analysis about its application or implementation. A summary of the criticisms of EIAs is made in Box 7.7.

The role of the private sector

Although governments may have a range of powers and planning measures at their disposal to protect the natural environment, the commitment of the private sector to environmental protection is essential to enhance chances of success. The extent to which private organisations decide to take initiatives of environmental protection will be dependent upon their philosophy, values and available resources. However, as the United Nations Environment Programme (UNEP) (2005) observes, there are advantages to be gained from the private sector playing an active role in conservation and environmental protection. These include: cost savings; preservation of their main business assets, i.e. the natural and cultural environments their customers are willing to pay to visit; and enhancement of their reputation as an environmentally responsible organisation. This latter point may be of special significance in the changing global context of how we think about and make ethical judgements upon our interaction with nature.

Box 7.7

Criticisms of environmental impact analysis (EIA)

- Costly to implement because of its requirement for a variety of different specialists
- Problem of avoiding bias originating from the ownership of the study
- The preparation of the EIA can cause delays in the planning process and the process has been criticised for providing inadequate opportunities for public involvement
- Much tourism development is small scale, thereby falling out of mandatory requirements for an EIA, which tend to be restricted to larger-scale schemes. Owing to its small scale, environmental damage is incremental, cumulative and less easy to predict.

As the major player in the tourism system, the private sector has a major influence in determining the extent to which impacts from tourism upon the environment will be either positive or negative. Consequently, a proactive approach to environmental management is of critical importance for natural resource conservation. Examples of initiatives taken by tourism operators to enhance their environmental operations include the Responsible Tourism Initiative and Tour Operators Initiative described in Boxes 7.8 and 7.9.

In the context of conserving natural resources and minimising pollution, both within their own organisation and tourism destinations, an important technique is environmental auditing and management systems.

Environmental auditing and environmental management systems

Explaining the process of environmental auditing, Goodall (1994: 656) comments:

Box 7.8

Association of Independent Tour Operators (AITO) and Responsible Tourism Initiative

The Association of Independent Tour Operators (AITO) is an amalgam of independent companies, specialising in particular holidays or destinations. A key part of their mission is to practise 'Responsible Tourism', as is reflected in this statement: 'We acknowledge that wherever a Tour Operator does business or sends clients it has a potential to do both harm and good, and we are aware that all too often in the past the harm has outweighed the good.'

A set of guidelines have been developed that emphasise the responsibilities of tour operators and other tourism stakeholders. They are:
- Protect the environment – its flora, fauna and landscape
- Respect local cultures – traditions, religions and built heritage
- Benefit local communities – both economically and socially
- Conserve natural resources – from office to destination
- Minimise pollution – through noise, waste disposal and congestion.

Source: Association of Independent Tour Operators (2007)

Box 7.9

Tour Operators' Initiative (TOI) for sustainable tourism development

In 2000, tour operators from different countries joined together in an initiative for sustainable tourism development, committing themselves to integrate sustainable principles into their business practices. Members include organisations operating at an international level, such as Accor, Thomas Cook and the German-based TUI Group. The initiative was developed with the support of the United Nations Environment Programme (UNEP); the United Nations Education, Scientific and Cultural Organisation (UNESCO) and the World Tourism Organization (WTO). Highlighting the link between environmental commitment and successful tourism, the TOI (2007) comment:

> It is in the tour operators' interest to preserve the environment in destinations and to establish good relations with local communities.... Tour operators have a central role in the tourism industry, acting as intermediaries between tourists and tourism service suppliers. They influence consumer demand, destination development patterns, and their suppliers' performance, as well as tourists' behaviour.

Three key areas have been prioritised for TOI action: supply chain management; co-operation with destinations; and sustainability Tour Operators' Initiative (2007).

> Environmental auditing provides the basis for such business practice [improving the current environmental performance of tourism firms] and is consistent with the view of management as a controlled cyclic process based on continuous monitoring of impacts and change, the development of knowledge and the feeding back of these into decision making by formalised process.

The reasons why businesses may be encouraged to participate in environmental auditing fall into three main categories. First, the passing of environmental legislation and enforcement of punitive measures against tourism firms who are polluters of the environment may encourage a company to seek to improve its environmental quality. Second, if companies believe they can reduce their costs of operations and increase their profits through the utilisation of environmental auditing, they are likely to pursue auditing as a course of action. Third, some companies may be genuinely philanthropic and willing to adopt as many measures as they can reasonably afford, to benefit the physical and social environments.

According to Parviainen et al. (1995), an environmental or eco-audit would typically cover aspects of environmental management, including: the company's environmental and purchasing policies; the adequacy of its communication of environmental practices to its staff and their level of environmental training; impacts of the business upon the surrounding physical environment, including features such as air, water, soil and ground water, noise and aesthetics; energy usage; and waste management and waste water schemes. They also point out that environmental audits form an integral part of a wider environmental management system (EMS) for businesses.

Environmental management systems integrate strategic objectives for the environmental quality of a company's operation with the practical aspects of environmental auditing. The first stage of an EMS is for a company to state clearly that it has an environmental commitment, which if taken seriously will subsequently influence the operations of the company. The next stage is to outline broad objectives of what the company hopes to achieve; for example, for a hotel one objective might be to reduce the amount of untreated waste emitted into the sea. The company would then carry out an eco-audit of its operations, determine realistic targets of what can be achieved within a certain timeframe, and develop mechanisms to achieve the targets. An essential part of the scheme is the ongoing monitoring of operations to determine if the targets set for environmental improvements are being met. If they are not, then strategies must be developed to rectify the situation.

Developing an EMS is a long-term commitment and is likely to take several years to incorporate all the different stages, from policy to review. The EMS system is not exclusive to any size of business but the resources available to any particular organisation will have an influence on the quality of the scheme. Importantly, it will require an investment of time and commitment from all employees of the organisation.

Within the EMS system, the eco-audit becomes a tool to evaluate the company's performance, and to make subsequent alterations to environmental policy and plans of action. The use of EMSs in the tourism industry is limited, yet it offers an approach for businesses that is both environmentally beneficial and proactive. The benefits to the industry of using EMSs include:

- Reduced risk of financial liability for environmental damage
- Improved customer relations
- Reduction in operating costs
- Improved access to lenders, insurers and investors
- Improving environmental resources without regulatory requirements and government interference.

(After Todd and Williams, 1996)

One scheme that demonstrates the advantages of a proactive stance by government towards influencing the relationship between tourism and the environment, and the advantages of a healthy public–private sector partnership, is that of the 'YSMEK' project in Finland (see Box 7.10). The aim of the scheme was to encourage tourism businesses to undertake environmental audits to gain a deeper understanding of their environmental practices, and also to help to develop a more sustainable

Box 7.10

Environmental auditing in Finland

In 1994 a pilot project was initiated by the Finnish Tourism Board to improve the environmental performance of tourism businesses in Finland. The scheme, known as the 'YSMEK' project, was funded by the Ministry of the Interior and the Ministry of Finance. YSMEK involved ten tourism organisations undertaking eco-audits. These included city hotels, farm tourism establishments, a ski resort, spa hotel, and training and exercise centres. The eco-audits concentrated primarily on the hotel and restaurant operations of the businesses, although for rural tourist enterprises, impacts upon agriculture and forestry were also included.

As a consequence of the environmental audits, the companies have reduced their usage of disposable products, decreased their consumption of raw materials, water and energy, and now produce less waste. Average cost savings have been 10 to 15 per cent for electricity and 30 per cent for water.

Source: Parviainen *et al.* (1995)

form of tourism in Finland. Importantly from the perspective of businesses, the environmental audits demonstrated that operational savings could be made by improving environmental practices.

One company that took the initiative over the impacts of its operations on the environment at the beginning of the 1990s (previously referred to in Box 7.9) is the large German-based tour group Touristik Union International (TUI). The use of environmental audits has now become a regular part of TUI's business operations (see Box 7.11).

Box 7.11

Touristik Union International (TUI)

Although the tour-operating industry can historically be criticised for generally taking little interest in environmental concerns, one notable exception, operating in the mass tourism market, is the German group Touristik Union International (TUI). In terms of volumes of sales, TUI is a major player in the European market.

TUI's interest in the environment reflects both concern about the impacts of mass tourism and a pragmatic business sense to respond to the demands of the German tourism market, where the environmental quality of destinations is known to have a critical influence on the level of customer satisfaction and subsequent demand. TUI has therefore realised that by investing in the protection and conservation of the environment, it is helping to safeguard its own financial future.

In 1990, TUI was the first tourism company to appoint an environmental manager as a member of its management board, and now has an established environmental unit in the company, dealing specifically with environmental matters. Apart from carrying out environmental audits of TUI's operations, the unit also consults and liaises with the following: the governments of host countries; international and national public organisations holding responsibility for tourism and the environment; regional and local authorities; their business partners, including hotels, airlines, car rental companies; and, importantly, their customers, to make them aware of good environmental practices.

The advantage of the TUI group in terms of influencing policy over the environment is that owing to its size (nine tour operators, five hotel companies with 130 hotels, and 700 travel agencies) it holds huge political sway. The TUI approach is innovative in the tourism industry because it incorporates environmental protection as a fundamental management function in the organisation of its companies, thereby fulfilling a goal of Agenda 21, which is discussed in the next chapter.

An example of a hotel chain that has taken the action to introduce an environmental management scheme into its operations is Grecotels, the largest hotel company in Greece. Initially, the scheme involved hotel auditing programmes, but this has now been extended to include the strengthening of links between Grecotels and local suppliers in a way that is not detrimental to the environment, for example, through the encouragement of organic agriculture for consumption in the hotels, as discussed in Box 7.12.

Box 7.12

Grecotels: environmental auditing and organic farming

Grecotels is the largest hotel company in Greece with 22 hotels and a bed capacity of approximately 11,000. In 1992 it was the first Mediterranean hotel group to form an environmental department, and initiated an environmental management and protection programme, initially supported by the European Commission. Between 1992 and 1993 professional eco-audits were conducted on the six hotels which at that time comprised the Grecotel chain. As a consequence of these audits, the following actions were taken, which are examples of the practices employed by the group in five different operational areas:

1 *Technical matters*
- Improved biological treatment plants and reuse of water for garden irrigation whenever possible
- Solar-heated water, up to 100 per cent in some hotels
- Wherever possible, sea water is used in the swimming pools
- Non-CFC fridges
- Flow regulators on all bathroom taps to reduce water wastage.

2 *Purchasing policy*
- Financing of organic agricultural production in Crete to supply vegetables and organic produce to the Grecotels situated there
- Recycling and reuse of waste products wherever possible
- Toxic and dangerous chemicals replaced with more eco-friendly products or reduced usage
- Promotion of Greek traditional local products in hotel shops and food outlets.

Box 7.12 continued

3 *Built environment*
- Use of local building materials in new hotels and for the renovations of older Grecotels
- Reintroduction of local plant species into hotel gardens.

4 *Communication to personnel and guests*
- Training of all personnel in the environmental policy of Grecotels and environmental duties are included in all job descriptions
- Production and distribution of an eco-leaflet to all personnel encouraging ecological thinking in the home as well as the workplace
- Results of programmes are given to guests and their active participation is encouraged in schemes
- Conveyance of the scheme and results to other hoteliers, tourism associations, tourism students and policy-makers.

5 *Cultural and social*
- Sponsorship of archaeological sites on Crete
- Co-operation with environmental organisations to protect endangered species and their habitats, for example, the Sea Turtle Protection in Rethymnon.

An interesting aspect of the Grecotel programme is that it demonstrates how a major hotel player can strengthen linkages to other economic sectors through its purchasing policy. Grecotels appointed a professional economist specialising in organic farming to add technical support to Grecotel's agricultural activities, and the company funds an organic farming project. The use of environmental procedures, besides benefiting the environment, has also proved to be financially beneficial to Grecotels. For example, the use of organic techniques to grow the garden flowers has reduced the dependency upon fertilisers and pesticides and organic waste is used for compost, making annual savings of tens of thousands of US dollars. Links to other sectors of agriculture have also been strengthened, including co-operation with Cretan vine growers to produce bottles of local wine for use in the hotels. An additional benefit of using local producers is that there is a saving in fuel consumption and reduction in air pollution due to the minimisation of transportation needs and 'food kilometres'. Olive oil, herbs and other local products are also promoted in Grecotels, providing a means for local families to generate income.

Is it all good news?

The case studies above illustrate the positive advantages for businesses from making environmental improvements in their procedures and practices. Importantly, from a business perspective, they demonstrate that through environmental audits, it is possible to reduce the operational costs of businesses. The role of government is also illustrated by the Finland case study as being important in aiding the implementation of auditing schemes in the sector. Government support is particularly important for the many small and medium-sized enterprises that constitute the bulk of the tourism industry, as these may want to be involved in environmental improvement schemes, but lack the technical expertise to be able to do so. The Grecotel project also shows how a major player can be of benefit to other sectors of the economy, in terms of stimulating demand for local suppliers, whilst helping to improve the environmental standards of agriculture and quality of food for guests.

There is now the evidence to suggest that environmental responsibility has been put on the agenda of many tourism organisations with varying degrees of importance attached to it. However, it is also the case that the entrenched attitude of much of the sector persists, in that until customers demand environmental improvements, companies will be unwilling to make environmental commitments, especially if they cost money and make inroads into their profit margins. There is also a threat that the 'greening' of tourism organisations becomes part of a superficial fashion, rather than representing a meaningful attempt to have a positive relationship with the natural environment. Consequently, how quickly an environmentally philanthropic approach spreads through the industry will depend upon the buying behaviour of consumers or tourists. Specifically, the extent to which they themselves are 'green', and willing to purchase on the basis of environmental policies and actions of companies, will be highly determinatory for pushing the green agenda. A further issue is the extent to which tourism organisations can pursue an environmental agenda without damaging their profits to an unacceptable level. As was explained in Chapter 4, the market has failed to cost environmental externalities into the prices of goods and services. Until this major deficiency is properly rectified, it is unlikely to be in companies' financial interests to maximise their environmental efficiency.

THINK POINT

Do the environmental credentials of the organisation you are buying a flight, hotel room or holiday from influence your purchasing decision? Explain the reasons why they do or do not.

Environmental codes of conduct for tourism

In attempts to guide the behaviour of organisations and individuals, voluntary codes of conduct to mitigate the negative impacts of tourism and improve environmental quality have been developed by government, the private sector and NGOs in the last few years. The usefulness of codes of conduct in tourism vis-à-vis other approaches to improve tourism's interaction with the environment is described by UNEP (1995: 3) as follows:

> A wide range of instruments can be used to put the tourism industry on the path to sustainability. Regulations, of course, are – and will remain – essential for defining the legal framework within which the private sector should operate and for establishing minimum standards and processes. Economic instruments are also being increasingly used by governments to address environmental issues. However, voluntary proactive approaches are certainly the best way of ensuring long-term commitments and improvements.

The primary aim of codes of conduct is to influence attitudes and modify behaviour (Mowforth and Munt, 1998). The objectives of codes of conduct for tourism are shown in Box 7.13.

These objectives cover a wide range of the stakeholders in tourism, including the private sector, government, local communities and tourists. Consequently, the developers of codes of conduct come from a wide variety of organisations, including governments and national tourist boards, the tourism industry and trade associations, and non-governmental organisations, such as the Ecotourism Society and the World Wide Fund for Nature. Codes tend to be targeted at the tourism industry, local communities involved with tourism, and tourists. According to Goodall and Stabler (1997), the salient principles of the codes relating to the tourism industry include:

- The sustainable use of resources
- Reduction of environmental impacts, for example, atmospheric emissions and the disposal of sewage
- Reducing waste and over consumption, for example, increasing the amount of recycling
- Showing sensitivity for wildlife and local culture
- Adopting internal environmental management strategies such as environmental auditing
- Support and involvement of the local economy by using local suppliers where possible
- Pursuing responsible marketing.

Box 7.13

The objectives of codes of conduct for tourism

- Serve as a catalyst for dialogue between government agencies, industry sectors, community interests, environmental and cultural NGOs and other stakeholders in tourism development
- Create an awareness within industry and governments of the importance of sound environmental policies and management, and encourage them to promote a quality environment and therefore a sustainable industry
- Heighten awareness among international and domestic visitors of the importance of appropriate behaviour with respect to both the natural and cultural environment they experience
- Sensitise host populations to the importance of environmental protection and the host–guest relationship
- Encourage co-operation amongst industry sectors, government agencies, host communities and NGOs to achieve the goals listed above.

Source: UNEP (1995: 8)

A typical example of a code of ethics developed by the industry is that of the Tourism Industry of Canada, as shown in Box 7.14.

Codes of conduct may also be established for local communities which are affected by tourism. Such codes can be helpful in the following ways: advising on the role of the local population in tourism development; safeguarding local cultures and traditions; educating the local population on the importance of maintaining a balance between conservation and economic development; and providing quality tourist products and experiences (UNEP, 1995). An example of a local community code is one developed by the non-governmental organisation, 'Tourism with Insight and Understanding', based in Germany. The code stresses the importance of community participation in tourism development, the need for respect of the local culture from tourists, and the need for tourism to play a balanced part in the economy of the region. An abridged version of the code is shown in Box 7.15.

The final type of code is aimed specifically at the behaviour of tourists. Broad guidelines produced in such codes usually include the following: learning as much as possible about your destination; using suppliers (such

Box 7.14

Code of ethics for the tourism industry – the Canadian example

The Canadian Tourism Industry recognises that the long-term sustainability of tourism in Canada depends on delivering a high-quality product and a continuing welcoming spirit among our employees and within our host communities. It depends as well on the wise use and conservation of our natural resources; the protection and enhancement of our environment; and the preservation of our cultural, historic and aesthetic resources. Accordingly, in our policies, plans, decisions and actions, we will:

1 Commit to excellence in the quality of tourism and hospitality experiences provided to our clients through a motivated and caring staff.
2 Encourage an appreciation of, and respect for, our natural, cultural and aesthetic heritage among our clients, staff and stakeholders, and within our communities.
3 Respect the values and aspirations of our host communities and strive to provide services and facilities in a manner which contributes to community identity, pride, aesthetics and the quality of life of residents.
4 Strive to achieve tourism development in a manner which harmonises economic objectives with the protection and enhancement of our natural, cultural and aesthetic heritage.
5 Be efficient in the use of all natural resources, manage waste in an environmentally responsible manner, and strive to eliminate or minimise pollution in all its forms.
6 Cooperate with our colleagues within the tourism industry and other industries, towards the goal of sustainable development and an improved quality of life for all Canadians.
7 Support tourists in their quest for a greater understanding and appreciation of nature and their neighbours in the global village. Work with and through national and international organisations to build a better world through tourism.

Source: Tourism Industry Association of Canada (1995)

Box 7.15

A code of conduct for host communities

1 Tourism should supplement our economy appropriately. We know, however, that it also represents a danger to our culture and our environment. We therefore want to supervise and control its development so that our country may be preserved as a viable economic, social and natural environment.

2 By independent decision making in tourism development we mean that the host population should decide on and participate in all matters relevant to the development of their region; tourist development by, with and for the local population. We encourage many forms of community participation without neglecting the interests of minorities.

3 The tourism development we aim for is economically productive, socially responsible and environment-conscious. We are prepared to cease pursuing further development where it leads to an intolerable burden for our population and environment.

Source: UNEP (1995; abridged version)

---

as airlines, tour operators, travel agents and hotels) that demonstrate a commitment to environmental practices; respecting local cultures and traditions; aiding local conservation efforts; supporting the local economy by buying local goods and services; and using resources in an efficient and unwasteful manner. A typical code of conduct for tourist behaviour is shown in Box 7.16, produced by the non-governmental organisation, Tourism Concern, for trekkers in the Himalayas.

Criticism of codes

Although the development of codes of conduct offers a way forward in making all the stakeholders in tourism aware of their environmental responsibilities, there have been several criticisms made of them. According to Mowforth and Munt (1998: 121), there exist a number of problems and issues arising with codes, including 'the monitoring and evaluation of codes of conduct; the conflict between codes as a form of

Box 7.16

The Himalayan Tourist Code

By following these simple guidelines, you can help preserve the unique environment and ancient cultures of the Himalayas.

Protect the natural environment

- **Limit deforestation – make no open fires** and discourage others from doing so on your behalf. Where water is heated by scarce firewood, use as little as possible. When possible choose accommodation that uses kerosene or fuel efficient wood stoves.
- **Remove litter, burn or bury paper** and carry out all non-degradable litter. Graffiti are permanent examples of environment pollution.
- **Keep local water clean and avoid using pollutants** such as detergents in streams or springs. If no toilet facilities are available, make sure you are at least 30 metres away from water sources, and bury or cover wastes.
- **Plants should be left to flourish in their natural environment** – taking cuttings, seeds and roots is illegal in many parts of the Himalayas.
- **Help your guides and porters to follow conservation measures.**
- **When taking photographs, respect privacy** – ask permission and use restraint.
- **Respect Holy places** – preserve what you have come to see, never touch or remove religious objects. Shoes should be removed when visiting temples.
- **Giving to children encourages begging.** A donation to a project, health centre or school is a more constructive way to help.
- **You will be accepted and welcomed if you follow local customs.** Use only your right hand for eating and greeting. Do not share cutlery or cups, etc. It is polite to use both hands when giving or receiving gifts.
- **Respect for local etiquette earns you respect** – loose, light-weight clothes are preferable to revealing shorts, skimpy tops and tight-fitting action wear. Hand holding or kissing in public are disliked by local people.
- **Observe standard food and bed charges** but do not condone overcharging. Remember when you're shopping that the bargains you buy may only be possible because of low income to others.
- **Visitors who value local traditions encourage local pride and maintain local cultures**, please help local people gain a realistic view of life in Western Countries.

The Himalayas may change you –
please do not change them.
As a guest, respect local traditions,
protect local cultures, maintain local pride.
Be patient, friendly and sensitive.
Remember – you are a guest.

Source: Tourism Concern

marketing and codes as genuine attempts to improve the practice of tourism; regulation or voluntary self-regulation of the industry; and the variability between codes and the need for coordination'.

There is an obvious concern that the codes produced by industry represent little more than a cynical marketing ploy and an attempt to stave off any kind of government intervention in terms of environmental regulation. The production of a code by a tour operator printed in a brochure, for example, may satisfy a consumer who is searching for a more ethical holiday, but may mean relatively little in terms of changed business practices by the operator. There is also the major question of whom, if anyone, will take responsibility for monitoring the effectiveness of the codes. This has meant there has been a lack of a general evaluation of codes as Mason and Mowforth (1996: 163) comment: 'There has been a clear lack of monitoring and evaluation of codes of conduct for the purpose of addressing their uptake and effectiveness.' Over ten years later, this situation has not changed.

In similar fashion, Goodall and Stabler (1997) talk of the limited practical usefulness of the codes because of their concentration upon principles, rather than informing tourist businesses on best environmental practice, and how this can be implemented in their own organisation. They also point out the spatial limitations of the vast majority of codes, which are destination based, and thus ignore the consequences of tourism in generating and transit areas.

Summary

- The role of government is critical in formulating environmental policy for tourism. This includes the establishment of protected areas, especially national parks, for the purposes of nature conservation and recreation. In some cases, particularly in developing countries, revenues from tourism will be used to fund park organisation and management. Through legislative and fiscal powers, it has a range of measures to encourage prioritisation of the environment by the different stakeholders involved in tourism, especially the tourism industry. The support of the tourism industry is essential for the success of a resource conservation-based model of tourism.
- A range of land-use planning and management techniques can be used to achieve environmental conservation. These include zoning; carrying

capacity analysis; limits of acceptable change; and environmental impact analysis. Environmental auditing and management systems are key managerial techniques that can be adopted by the tourism industry. Not only can they improve environmental quality, but they can also reduce companies' operating costs, besides improving their public image. However, for the many small and medium-sized enterprises that constitute the bulk of the tourism industry, technical assistance will be needed to help them implement environmental improvements.

- Although notable initiatives have been taken within the tourism industry towards environmental responsibility, e.g. the Tour Operators' Initiative, the influence of the consumer and market will be central in directing the industry. The influence of the consumer will be dependent upon the extent to which a company's environmental credentials and record are seen to influence purchasing behaviour. The market will also influence the propensity for organisations to accept environmental responsibilities if negative externalities are reflected in the costs of tourism goods and services.

- Codes of conduct can be developed and adapted by different stakeholders in tourism. For the private sector they can be interpreted as a form of self-regulation rather than regulation being imposed by government. However, criticisms can also be made of codes of conduct, including lack of monitoring and evaluation of their usefulness, and spatial limitations.

Further reading

Eagles, P.F.J., McCool, S.F. and Haynes, C.D. (2002) *Sustainable Tourism in Protected Areas: Guidelines for Planning and Management,* Cambridge: IUCN.

UNEP (2005) *Integrating Sustainability into Business: A Management Guide for Responsible Tour Operations,* Paris: United Nations Environment Program.

Suggested websites

Responsible Travel: www.responsibletravel.com

United Nations Environment Program: www.unep.org

US Environmental Protection Agency: www.epa.gov

8 Climate change, natural disasters and tourism

- Understand the causes and significance of climate change
- Consider the reciprocal relationship between climate change and tourism
- Explain the meaning of natural disasters
- Assess the impact of natural disasters upon tourism

Introduction

At a time when the debate on global warming and associated climate change has gained prominence in international and national policy, academia and the media, primarily because of its threat to our economic and social well-being, it is relevant to discuss the relationship that tourism has with it, as both an instigator and recipient of climatic change.

A theme of this book is that tourism does not take place in a void without impacting upon and being impacted upon by the natural environment. Another theme is that tourism is heavily dependent upon climate and natural resources. Climate, beaches, oceans, mountains, forests, wildlife and their associated ecosystems provide the attractions for many destinations. Supplies of natural resources such as fresh water are essential to keep tourism destinations functioning. Upon these resources, regular international and domestic flows of tourism have been constructed over decades.

Established flows of tourism are usually reliant upon certain assumptions, for example, that the Mediterranean will have a warm and dry summer, the Caribbean will have its dry season at a predictable time each year, and the mountain ski resorts of North America and Europe will have regular snowfall in winter. However, it would seem that these assumptions are now under threat from predicted climatic changes as a consequence of global warming. Worryingly for those whose livelihoods are dependent upon tourism, there is uncertainty over how climate change will affect tourism demand.

The significance of climate change

The terms 'global warming' and 'climate change' have become incorporated into the global vernacular of international debate and conversation. Scientific evidence confirms that global warming is occurring (IPCC, 2001, 2007a, 2007b), and the experience of extreme or unusual weather patterns by people in many parts of the world would seem to support the scientific evidence. A limited amount of contention may still exist over the extent to which global warming is a consequence of human activity or a natural occurrence, but the key debate has now shifted to whose responsibility it is to deal with its causes and mitigate its impacts.

Although numerous people may be able to give anecdotal accounts of experiencing strange weather, it is important to make a distinction between 'weather' and 'climate'. Whilst our weather may change daily, climate is defined as the: 'combination of weather conditions at a particular place over a period of time – usually a minimum of 30 years' (Collins, *Dictionary of Geography*, 2004: 73). There is subsequently a long-term dimension to climate that involves the monitoring of weather conditions.

The primary influences upon climate are degrees of latitude; ocean currents; the movements of wind belts and air masses; temperature differences between land and sea surfaces; and vegetation. Given that climate is a combination of weather events over a minimum of three decades, 'climatic change' has a longitudinal dimension and is a consequence of variations in some or all of the preceding factors over variable periods of time. For example, a variation in temperature differences between land and sea-surfaces may occur in decades as a consequence of global warming, whereas latitude shifts in land mass as a process of plate tectonics takes tens of millions of years. As climate is a combination of weather events over a minimum of three decades, the meaning of climate change can be defined as:

> Change in the climate of an area or the whole world over an
> appreciable period of time. That is, a single winter that is colder than

average does not indicate climate change. It is the change in average weather conditions from one period of time (30–50 years) to the next.

(Ibid.: 75)

The Earth's climatic diversity has in turn produced a variety of biodiverse ecosystems, e.g. rainforests, deserts, coral reefs and mangroves, many of which offer resources for tourism, as has been discussed in previous chapters. Critically, many of these ecosystems are reliant upon the consistency of the climatic conditions that created them, and are consequently vulnerable to climatic change, as was explained with the example of coral reefs in Chapter 3. The threat to the Earth's biodiversity, including the well-being of humans, is a major concern of climate change.

Throughout the history of the Earth, its climate has been subject to fluctuations and changes. Of major importance in determining the patterns of the global climate is the relationship between the quantities of energy received from the sun, vis-à-vis the amount of energy lost from the Earth. The amount of energy received by the Earth from the Sun is influenced by a number of factors. These include natural variations, notably fluctuations in solar activity; small variations in the Earth's orbit around the Sun (Milankovitch cycles); and any process that influences the amount of particulate matter in the atmosphere, for example, volcanic activity or an asteroid impact will increase the amount of particulate matter in the atmosphere, reducing the amount of solar energy reaching the surface (Stern Report, 2006). Nor is the distribution of the Sun's energy uniform across the Earth's curved surface, with the uneven fall of the Sun's parallel rays causing more heat to be received at the Equator than at the poles, the major factor in explaining why temperatures at the Equator are considerably higher than those at the poles.

It is these variations that account for the natural fluctuation in the Earth's global temperature which lead to mini ice ages and hotter periods. However, any changes in the global temperature that cause climate change have traditionally occurred over long periods of time, i.e. thousands of years. This is very different to the rate of temperature increase that is being witnessed today. In the last century, the global mean air temperature close to the surface rose by 0.6 degrees Celsius but it is in the latter half of the century that the trend was most evident. The rate of the increase is exceeding anything that has happened in the past 20,000 years, whilst the present levels of CO_2 concentration have not been exceeded in the last 420,000 years (IPCC, 2001). Eleven of the last 12 years (1995-2006) rank amongst the 12 warmest since 1850,which marked the beginning of meteorological records, with 1998 and 2005 being the hottest. The 1990s marked the warmest decade, a record that is highly likely to be surpassed by the first decade of this century (Grassl, 2006; IPCC, 2007b).

At the same time the average global climatic temperature has been rising, there has been a positive correlation with increased levels of CO_2 in the atmosphere. Carbon dioxide forms one of what are referred to as the greenhouse gases (GHGs), which possess a chemical structure that enables them to absorb and re-emit heat. They include water vapour (H_2O); carbon dioxide (CO_2); methane (CH_4); nitrous oxide (N_2O); ozone (O_3) and chlorofluorocarbons (CFCs) (Stern Report, 2006). All of these perform a crucial function in maintaining the Earth's climate by trapping heat at its surface. They also possess different heat-retaining properties, for example, methane is 23 times more efficient at trapping heat than CO_2, whilst nitrous oxide is 296 times more efficient. Consequently, the chemical balance of the atmosphere is critical in influencing the amount of energy in the form of heat that is retained close to the Earth's surface. As the concentration of GHGs in the atmosphere increases, by implication the Earth's temperature is likely to increase.

The atmospheric level of CO_2 has been constant at 280 parts per million (ppm) of atmosphere for the last 1,000 years. However, since the commencement of the Industrial Revolution at the end of the late eighteenth century, the level has risen to 380 ppm today (Stern Report, 2006). This rise is associated with the release of CO_2 into the atmosphere from the burning of natural carbon stocks, notably oil and coal. However, CO_2 is not the only GHG to have been released into the atmosphere. Methane releases are also significant, leading scientists to conclude that equivalent levels of CO_2 in the atmosphere, i.e. CO_2e, stand at 430 parts per million (ppm) (De Costa, 2006).

It would seem that there is a positive correlation between industrial development and increased levels of CO_2e in the atmosphere. The source of the majority of CO_2e emissions has been from the developed countries, especially North America and Europe. Since 1850, North America and Europe have produced around 70 per cent of all the world's CO_2 emissions due to energy production (Stern Report, 2006). However, this situation is rapidly changing, notably with the growth of China and India as global economic powers. It is expected that China would have overtaken the US as the world's biggest emitter of GHG by 2010. According to the International Energy Agency (IEA), if China takes no action to restrain CO_2 emissions, at the current growth rate it would be emitting twice as much CO_2 within 25 years as the world's 26 richest countries put together (Vidal, 2007).

Nevertheless, even if global CO_2 emissions were held at their present level, global warming and the heating of the Earth's surface and oceans would continue (European Environment Agency, 2006). This is because the efficiency of the Earth's system to remove CO_2 from the atmosphere is slower than the present rates of addition, whilst the oceans absorb 80 per cent of the heat added to the climatic system (IPCC, 2007b). They would reach 550 ppm by 2050, which is double their pre-industrial levels (Stern Report, 2006). Such an increase would lead to a high probability of exceeding an increase of 2 degrees centigrade in global average temperature. If no action to reduce CO_2 emissions is taken in a Business-as-Usual (BAU) scenario, by the end of the twenty-first century, the stock of greenhouse gases could treble, giving at least a 50 per cent probability of exceeding a 5 degrees centigrade increase in global average temperature (ibid.).

To give an indication of what this range of predictions mean in practical terms for the climate: the Earth's global average temperature is only 5 degrees centigrade higher than at the time of the last ice age. Although it is impossible to predict the consequences of climate change with complete certainty, not least because of the uncertainty of the extent to which the global surface temperature will actually increase, there will be impacts upon natural resources and ecosystems. For example, within a range of 1 to 5 degrees centigrade increase in global average temperature these include (to varying levels of intensity): significant extinctions of species; widespread coral mortality; ecosystem changes; a decline in mountain glaciers, snow cover and ice sheets leading to sea-level rises; acidification of oceans as a consequence of increasing atmospheric levels of CO_2; significant loss of biodiversity; hundreds of millions of people exposed to increased water stress; increased damage from floods and storms; up to 30 per cent of global coastal wetlands lost; millions more people exposed to coastal flooding; and increased mortality from drought, disease and malnutrition (IPCC, 2007a, 2007b).

As stated at the beginning of the chapter, although global warming and climate change have been highly contentious issues, there is now a fairly unanimous agreement in the scientific and political communities that they are both occurring. However, how we choose to respond to global warming will be influenced by a number of factors, including:

Our level of awareness of the specific environmental problems, our perception of the scale of the threat, whether the impacts are direct or indirect, whether the impacts directly affect the individual under consideration, and which groups in society and which countries suffer most.

(Belle and Bramwell, 2005: 33)

The influence of climate change on tourism

The uncertainty of the science of climate change prediction means there is no absolute certainty about how tourism will be affected by climate change. However, melting ice caps, rising sea-levels, reduced snowfall, loss of biodiversity and changing ecosystems will inevitably have implications for tourism. Owing to the importance of weather and natural environments to tourism, it is one of the sectors of the economy along with agriculture that is most likely to be affected by climate change. According to the World Tourism Organization (2003), specific threats include: that sea-levels will rise, threatening many coastal areas and small islands; temperature rises will change precipitation patterns, water supply problems will be exacerbated; and climate change will increase the magnitude, frequency and risk of extreme climatic events such as storms and sea surges. More specifically, it is probable that as the sea-level rises, there will be increased beach and coast erosion; a higher likelihood of coastal flooding; loss of coastal ecosystems; and a total submersion of some low-lying islands and coastal plains.

The possible effects of climate change on a typical small island developing state (SID) can be exemplified by Barbados in the Caribbean. Based on a study of the south and west coasts of the island to estimate the effects of a 1 metre rise in sea-level and a storm surge generated by a Category 3 hurricane, the Caribbean Planning for Adaptation to Climate Change (CPACC) agency, cited in Belle and Bramwell (2005), commented that: 'the result is astonishing since most of the present day development, including the tourism infrastructure, is located within this inundation zone' (CPACC, 1999: 3). A further potential consequence of a rise in sea-level is the intrusion of saline water into the fresh water aquifers that Barbados is dependent upon for water. Potential rises in sea-levels pose real threats to the tourism industries of low-lying islands such as the Maldives and Seychelles.

Alongside the threats climate change poses for small islands and coastal tourism, the continuation of winter sports will also be threatened by less reliable and infrequent snowfalls, especially in ski resorts at lower-level altitudes. There may be a subsequent displacement of demand to higher-altitude resorts, potentially placing them under increased environmental pressure. It is estimated that for every 1 degree centigrade increase in temperature, the snowline will rise 150 metres. The potential threat to ski resorts in the European Alps and Scotland from climate change is summarised in Boxes 8.1 and 8.2.

Box 8.1

Climate change: implications for the ski industry of the Alps

The Organisation for Economic Co-operation and Development (OECD) predicts that within 20 years in the European Alps, ski resorts below 1,050 metres will not be viable, and that by the end of the twenty-first century only the highest over 2,000 metres will be able to offer guaranteed snow. Similarly, scientists from the University of Zurich predict that by 2030, 50 of Switzerland's 230 ski resorts will not have enough regular snow to sustain skiing. Lower-altitude mountain villages in the central and eastern parts of Austria are also under threat.

Alongside the loss of enjoyment by the skier, the economic and social impacts upon these resorts would be dramatic. Given that many of them have limited economic diversity and are highly dependent upon tourism, with winter sports playing a major part, many livelihoods would be threatened. Businesses would face bankruptcy unless alternative markets based upon other types of tourism could be found. The linkages of tourism businesses to other parts of the economy would also be affected, threatening the livelihoods of those supplying services and goods to the tourism industry. Any displacement of demand to higher-altitude ski resorts would increase the environmental pressures upon the surrounding ecosystems. The reduced supply of downhill ski resorts, coupled with an existing high demand for the activity, would be likely to lead to a dramatic rise in prices, returning it to the elitist activity it once was.

Source: After Smith (2007)

Box 8.2

Climate threat: Cairngorms, Scotland

Situated in the north-east of Scotland, the Cairngorms are a massive granite plateau created during the last Ice Age. Its ecosystem is characterised by Arctic–Alpine fauna and flora and it is frequently referred to as a 'wilderness'. During the twentieth century it suffered depopulation, as younger people migrated to cities in search of employment, creating a social imbalance in communities. This trend was reversed during the last three decades of the twentieth century, owing to the development of a tourism industry based upon recreational activities and enjoyment of the qualities of the natural environment. To combat problems of seasonality, an important component of the tourism industry is winter sports, especially downhill skiing.

Climatic change in this area is already affecting animal, plant and human life. Skiing as a recreational activity has become increasingly marginal owing to infrequent snowfall, emphasised by Mr Bob Kinnaird, chairman of the Cairngorm Mountain Company that runs the ski area: 'Scotland's skiing prospects are not too good for the long-term future.' Native species of flora and fauna are being pushed further up the mountain as the amount of snow and number of snow-days decrease. However, as the climate changes, there will come a point where they can go no higher.

An indicative sign of climatic change is the decline in numbers of the bird called the snow-bunting, which breeds only in the Cairngorms and the Arctic. They like to make their nests on the edge of snowfields where they can eat larvae trapped in the snow. In 2005, 27 males were recorded, approximately two-thirds of the 1991 count. Estimates for 2006 are that they are down to half their 1991 count. Snow is also important for acting as a blanket that keeps the ground relatively warm and moist. As this protection disappears, the soil becomes exposed to freezing winds that bring wild temperature fluctuations, resulting in soil microbes being killed and a degradation of soil quality.

Source: McKie (2006)

As is suggested in Box 8.1, there will be implications from climate change for communities that rely upon tourism as an economic activity. Reduced demand will not only directly affect the livelihoods of those employed in tourism but also has implications for industries that have linkages to tourism, e.g. agriculture, fisheries, construction and handicrafts. Primary economic sectors such as agriculture and fishing will themselves also be affected by climate change.

However, whilst climate change may present threats to demand to some areas, for others it will present opportunities. For example, climate change may result in northern Europe becoming more attractive for summer vacations as the weather becomes consistently warmer and drier, whilst the Mediterranean may lose its appeal as temperatures become too hot, the landscape becomes more arid, and there is an increased frequency of natural disasters such as flash floods and forest fires (WTO, 2003).

In the case of tourism from North America to the Caribbean, a similar pattern of influences may affect travel. Similar to the Mediterranean, tourism in the Caribbean is primarily dependent upon climate and the beach, with its main market from North America escaping the cold winter climate. However, parts of the USA may become warmer, making them more attractive to vacation in, whilst rising sea-levels may threaten some of the Caribbean islands, damaging beaches and causing infrastructure damage (ibid.).

THINK POINT

Give examples of how climate change will influence tourism demand. What are the possible impacts upon peoples that are economically dependent upon tourism?

A predicted increased need for air-conditioning will also place pressure on the island's water and energy resources.

The impact of climate change may also be felt on special activity tourism besides more mainstream winter and summer tourism. As was explained in Chapter 3, coral reefs, the second most biodiverse ecosystem on the planet, are very susceptible to an increase in sea temperature, resulting in coral bleaching and the death of the reef. They are also vulnerable to increased hurricane and storm activity. The biodiversity of the reefs in many areas of the world attracts a lucrative diving niche market. The death of the reefs would place this under threat. Other types of tourism, also dependent on a stable climatic and ecosystems, e.g. wildlife tourism, could also be threatened by ecosystem change and the loss of biodiversity.

Tourism's contribution to climate change

As was discussed in Chapter 3, tourism has impacts and consequences upon the natural environment. Subsequently, whilst tourism will be affected by climate change, tourism itself is thought to be an inducer of change. Alongside pollution from tourism superstructure, e.g. hotels, attractions and restaurants, a major issue relates to the contribution of the transport element of tourism to global warming. For example, it is estimated that in the case of the USA, 76.5 per cent of tourism's contributions to greenhouse emissions is derived from transport, with the remainder coming from other tourist services, e.g. accommodation and restaurants (WTO, 2003).

Much of the debate about the contribution of tourism to global warming is centred upon the role of air transport, even though the majority of transport for tourism is based upon the car. However, it is the rapid growth in demand for air transport, combined with its high level of pollution per passenger kilometre compared to other types of transport, which causes particular concern, with some climatologists saying that aviation is the fastest-growing cause of climate change (Garman, 2006).

Although statistics date quickly, at the time of writing, global take-offs surpassed 2.5 million per month for the first time ever (McCarthy, 2007). Whilst the contribution of aviation to global emissions is relatively low, the amount of carbon dioxide pollution from aircraft is predicted to reach between 1.2 billion to 1.4. billion tonnes by 2025, compared to approximately 600 million tonnes now (Adam, 2007). As aircraft emissions are released at altitude, the global warming effect of CO_2 emissions is thought to be approximately 2.5 times higher compared to emissions from cars or power stations (ibid.).

Flying, in particular, has received high levels of media attention, partly because its association with tourism lends it to accusations of being an elitist and privileged activity. There is unlikely to be a technological solution to the problem, as the increases in demand and growth of flying mean that total aviation emissions continue to increase, despite improvements in emission-reducing technology (IPCC, 1999; Adams, 2007). The ultimate solution rests with ethical and economic principles. The ethical debate rests upon the willingness of individuals to forgo the personal pleasures gained from flying for the greater good of the natural environment and future generations. The economic principle rests with the price of airfares reflecting full environmental costs. If environmental taxes were placed on

THINK POINT

Accepting that pollution from aviation will play an increasingly significant role in global warming if demand continues to increase, is it ethically correct to fly to foreign destinations?

airfares to incorporate the costs of negative externalities caused by flying, there would be a likely decrease in demand, given the high price elasticity of the recreational tourism market.

Natural disasters and tourism

Global warming models predict that increasing global temperatures are likely to impact upon many atmospheric parameters, including precipitation and wind velocity, leading to more frequent extreme weather events, including storms, heavy rainfall, cyclones and drought (UNEP, 2000). The numbers of extreme weather events have quadrupled since the 1950s, and the frequency of high-level hurricanes and typhoons has doubled since the 1950s (De Costa, 2006). Combined with rising sea-levels and coastal erosion, there is a real threat from 'natural' or climatic disasters to tourism. Whilst these have always existed, they are likely to become more common.

According to UNEP (2000), natural disasters include earthquakes, volcanic eruptions, fires, floods, hurricanes, tropical storms, cyclones, landslides and other events that cause loss of life and livelihoods. As can be seen from this list, some natural disasters such as earthquakes, volcanoes and tsunamis are a consequence of seismic activity. However, the others are a product of climate, and therefore are likely to be influenced in their frequency by climate change.

Natural disasters have economic and social impacts, and it is estimated that almost three million people have perished as a result of natural disasters in the past three decades, while tens of millions have suffered hardship (UN, 1997). The most significant natural disaster of recent times in terms of lives lost was the Indian Ocean tsunami on 26 December 2004. It struck the coasts of 12 countries, killing approximately 200,000 (Goodwin, 2005). Another natural disaster of significance was the flooding of New Orleans in the USA as a consequence of Hurricane Katrina in 2005, which resulted in a mass evacuation of the city and led to thousands of people being homeless.

Hurricane Katrina was poignant because it was a natural disaster that struck at the most powerful and economically developed country in the world, demonstrating that it is not only the poor who are vulnerable to the power of nature. However, it is the poor who are most likely to suffer from major natural disasters, as was the case in the Indian Ocean. This is because the poor are often forced to live in marginal environments for habitation, e.g. land that is prone to flooding or landslides, and also because the construction of their homes is often of a poor quality, making them less able to withstand extreme natural events. Worryingly, Hurricane Katrina demonstrated that even the richest nation on earth struggled to cope with a natural disaster of high magnitude.

As was demonstrated with the Indian Ocean tsunami, the impacts of natural disasters may sometimes be international as well as regional or local. Some have a global impact, e.g. large volcanic eruptions can spread particles into the atmosphere around the whole globe, blocking sunlight and causing abnormal climatic conditions.

The environments of tourism, notably coastal areas and mountains, make them especially vulnerable to natural disasters. This vulnerability may be compounded by geographical location. For example, many tropical islands are characterised by intense climatic events such as hurricanes; the Himalayas are highly vulnerable to landslides and avalanches because of seismic activity; and as Belle and Bramwell (2005: 33) observe: 'Small island developing states (SIDS) are often especially susceptible to sea-level rise because they have long coastlines relative to land area and because large proportions of their area are low lying.' The potential negative impacts of natural disasters in many coastal areas are increased by the high density of building, including tourism infrastructure.

In destination areas, the combination of a downturn in tourism demand and damage to infrastructure makes the small and medium-size enterprises (SMEs) that dominate the industry highly susceptible to natural disasters. A major problem for SMEs is that often they do not have insurance cover, or the resources, to aid the rebuilding of their industry. Consequently, the livelihoods of those who work directly in the tourism industry and those who provide services and goods for it will be placed under threat. In tourism-generating areas, i.e. the places tourists come from, there will be a reduction in demand for travel to disaster-prone areas. The combination of media images, combined with perceived or actual risks, can be powerful in deterring tourists from travelling to a particular destination. The financial impact of this will be felt by tour operators and other tourism services.

However, the impact of natural disasters on the tourism industry and the responses of the industry have received little systematic research (Faulkner, 2001). Yet, given the location of many tourism destinations, it is probable that they will be at increased risk from natural disasters as a consequence of global warming, in the future. The kinds of impacts that a natural disaster may have on tourism are described in Box 8.3.

Box 8.3

The Indian Ocean tsunami and tourism

The Indian Ocean tsunami on 26 December 2004 graphically illustrates the threat of natural disasters to tourism destinations and the tourism industry. Triggered by an earthquake off the coast of north Sumatra, the tsunami struck the coastlines of 12 countries in the Indian Ocean, killing over 200,000 people.

Alongside the tremendous personal loss and psychological trauma, the tsunami had a significant impact on the economy of the region and liveli-hoods. Two countries badly affected by the tsunami were Sri Lanka and the Maldives. In Sri Lanka, the tsunami caused over 30,000 deaths and dis-placed 860,000 people (Christoplos, 2006). The costs to tourism-related industries on the island were estimated to be US$250 million and 27,000 tourism-related jobs were lost. In the case of the Maldives where tourism contributes approximately 33 per cent to the GDP directly and 60 to 70 per cent indirectly, the direct damage to the tourism infrastructure and other related businesses was estimated at over US$100 million (World Bank, 2005). The reality behind these figures is that for many people, they no longer had businesses to operate or tourists to serve, causing immediate cashflow problems and placing livelihoods at risk. Many of the SMEs in both places did not have insurance to cover the costs of rebuilding their businesses. The lack of insurance cover is a major issue in many natural disasters when finance is needed to restart businesses.

However, the recovery of the tourism infrastructure in many of the coun-tries affected by the tsunami within six months has been remarkable. Writing six months after the disaster, Bowes (2005:4) comments: 'With the help of the rest of the world, the area has gradually put itself back on track, with almost all affected areas now safe for tourists to visit.' However, tourist demand had virtually collapsed. In the case of Sri Lanka, of 48 of the 248 star-rated hotels that were damaged, six months later, 31 had been repaired.

Box 8.3 continued

In Phuket, Thailand, 420 SMEs closed for good, although the rest of the town was back to normal. A spokeswoman for the tourist board commented: 'Phuket is totally back to normal but it's like a ghost town.' Across Thailand, visitor numbers were 20 per cent down, whilst room rates were down 60 per cent. In the Maldives just 12 of the 87 resorts remained closed whilst visitor numbers were 30 per cent less than the previous year.

Source: After Bowes (2005) and Christoplos (2006)

Tourism's response to natural disasters

Given that the incidence of natural disasters affecting tourism is likely to increase in the future, as is the economic significance of tourism in many regions of the world, there is a need for a coherent response from the tourism sector to risks from natural disasters. As Faulkner (2001) suggests, it is inevitable that at some point there is a near certainty of tourism organisations experiencing a natural disaster.

Two potential responses include strategies and management to reduce the impacts (mitigation), and strategies to adjust to impacts (adaptation). As Faulkner suggests, any tourism disaster strategy for a destination requires a co-ordinated approach, with a designated tourism disaster management plan being in place that is representative of the private and public sector organisations involved in tourism. The voice of the 'community' should also be added to the disaster management plan, although achieving a truly cohesive opinion may be problematic. However, as was illustrated in the response to the Indian Ocean tsunami, no matter how good the internal organisation of a destination or a tourism business, the need for external help may be required.

Given that tourism destinations are highly reliant on their image for creating demand, a priority for any tourism destination aiming to recover from a natural disaster is to rebuild its image into one that appeals again to the market. Alongside the need to emphasise the 'beauty' of its natural and cultural resources, the requirement to emphasise the security for potential tourists and the minimisation of risk will be imperative. Faulkner (2001) suggests that a media communication strategy to ensure that misleading and contradictory information is not disseminated is an essential part of any tourism disaster management plan. Certainly, a negative image of a destination portrayed in the media as a consequence of a natural disaster will lead to a downturn in demand. Such an image will often linger longer than the time it takes to restore the general tourism infrastructure, as was the case for many of the affected areas of the Indian Ocean tsunami. Yet, whilst image is important, it must also be balanced against the reality of having a tourism infrastructure for tourists to go to, otherwise any recovery in demand is likely to be short-lived and the damage more long term.

Summary

- Levels of CO_2e in the atmosphere are increasing at a faster rate than at any time in the Earth's history. These greenhouse emissions, primarily caused by the burning of carbon stocks, i.e. coal and oil, are causing increases in the global average surface temperature. North America and Europe have been the main sources of these emissions but China is about to become the largest emitter of CO_2 emissions as its economy continues to expand rapidly. Eleven of the last 12 years rank amongst the warmest recorded since meteorological records began over 200 years ago.
- Predicting the effects of climate change is far from an exact science. Consequently, uncertainty exists how tourism will be affected by climate change. However, recreational tourism is dependent upon predictable climatic conditions, e.g. warm and dry summers; and snowy winters. If this dependency is threatened, then tourism demand will fall. Predicted rises in sea-levels will be problematic for small islands and coastal areas. However, for other areas climate change may present opportunities to secure a larger share of the tourism market. For example, northern Europe may become more attractive for summer vacations as the weather becomes consistently warmer and drier, whilst the Mediterranean may lose its appeal as temperatures become too hot, the landscape becomes more arid, and there is an increased frequency of natural disasters such as flash floods and forest fires. However, whilst tourism is threatened by climate change, it is also a contributor to global warming, notably through aviation emissions.
- Climate change is likely to lead to an increase in the frequency of natural disasters. Combined with rising sea-levels, tourism destinations in coastal areas and on small islands are likely to be places at

increased risk of damage. There is a subsequent need for disaster management planning for tourism to mitigate and adapt to these threats.

Further reading

Gossling, S. and Hall, M. (eds) (2006) *Tourism and Global Environmental Change: Ecological, Social, Economic and Political Interrelationships,* London: Routledge.

Intergovernmental Panel on Climate Change (2007) *Climate Change 2007: Impacts, Adaptation and Vulnerability*, Geneva: IPCC.

Stern Report (2006) *Stern Review on the Economics of Climate Change*, London: HM Treasury Office.

9 The future of tourism's relationship with the environment

- The growth of green consumerism and its effect upon tourism
- The significance of alternative tourism and ecotourism
- The future of tourism's relationship with the environment

Introduction

The concept of viewing tourism as a system linking the environments tourists come from with ones they go to has been emphasised as a central theme for considering tourism's relationship with the environment. The world now has a level of interconnectivity and reliability that means that many decisions made at a local level have a global impact. It is also evident that tourism's relationship with the environment can be either harmful or beneficial, and that perspectives on this relationship will involve ethical and value judgements alongside scientific evidence. This final chapter considers recent and emergent trends in tourism and society that are likely to influence the future of the tourism and environment relationship.

The growth of 'green' consumerism

Given that tourism can be understood as a type of consumerism, the extent to which environmental considerations influence buying behaviour is likely to have a major influence on how tourism interacts with the natural environment. By the end of the twentieth century, 'green consumerism' and the purchasing of 'environmentally friendly' products had become a significant niche market, a trend that has continued and strengthened in the first decade of the twenty-first century. Cairncross (1991) takes the view that the extensive media coverage of environmental concerns, such as climatic change, the burning of rainforests and the vanishing ozone layer, was highly influential in getting people to think about the environmental effects of what they were buying. Economic prosperity in developed countries has also been influential in moving people's concerns on from basic economic needs to wider ethical considerations of consumerism. Traditionally, levels of public concern over the environment tend to reflect economic cycles; that is, when the economy is buoyant, environmental concern is at its highest (Martin, 1997). However, whether environmental concern can continue to be an after thought of affluence, remaining otherwise unlinked to the market system, is doubtful if we wish to avoid the most dramatic aspects of global warming discussed in the last chapter. Nearly 20 years after Cairncross's observations, global warming and climate change have become integral issues of global culture, and are likely to have a significant impact upon the future production and consumption of goods and services.

Concern over the environment has manifested itself in consumer behaviour, in various ways, in different countries. For example, in Britain there was a virtual consumer boycott of aerosols that contained chloro-fluorocarbons (CFCs), after the media and environmental pressure groups alerted consumers to the role of CFCs in ozone depletion (Swarbrooke and Horner, 1999). In America, consumers' primary concern was with waste and excessive packaging on products, and German consumers' concerns also rested with packaging and the plastics used to make drink bottles (Cairncross, 1991).

This 'outbreak' of green consumerism forced companies to begin to respond to the environmental concerns of an affluent and sizeable minority of the market. Not only did this influence the retail side of the market but also the supply side of the market, as a few retailers began to check the environmental practices of their suppliers. For instance, the British hardware chain B&Q questioned their suppliers about the quality of their environmental management of the peat bogs from which the peat sold as fertiliser in their shops came from, and one supplier was subsequently discontinued for having poor environmental standards. In America, the three leading tuna canners, controlling 70 per cent of the market, decided to ensure that their fish were caught only in 'friendly ways' which were not harmful to other types of fish (Cairncross, 1991).

Other new companies emerged into the consumer market, based upon a philosophy of selling ethically sound and environmentally friendly products. One of the most notable business success stories has been The Body Shop, which now has hundreds of stores worldwide. The Body Shop's primary concerns are with skin and hair care, and it has emphasised from its foundation that environmental and social considerations have been at the forefront of its decision-making. The view of The Body Shop is made clear in its policy document: 'The Body Shop has always had a clear, top-level commitment to environmental and social excellence. Because of this, we make sure we include environmental issues in every area of our operations' (The Body Shop, 1992: 5). As part of this policy, the company does not carry out any animal experiments to test its cosmetics and also emphasises 'fair trade' practices with its suppliers, based upon buying from businesses who wish to protect their culture and practise traditional and sustainable land use.

There has also been a growing consumer demand for ethical investment funds, which can be traced to the early 1980s and the desire of the universities and churches in America not to put money into companies trading with South Africa, whose government supported the system of apartheid at that time (Cairncross, 1991). There has been a similar demonstration from some international organisations operating in the tourism sector of their environmental and green credentials, as was discussed in Chapter 7, partly influenced by a desire to be included in ethical investment funds listed on the global stock market.

The possible consequences for companies that ignore consumers' environmental concerns is illustrated by the case of Royal Dutch Shell, the world's biggest oil company. In 1995, Shell intended to bury an obsolete oil rig, called the Brent Spar, at the bottom of the North Sea. Concerns raised by Greenpeace, the non-governmental organisation, over the possibilities that toxic sludge released from the rig would contaminate the sea floor, led to a great deal of controversy over the plan. Activists encouraged boycotts of Shell service stations in Germany and motorists also began to shun Shell garages in Denmark and Holland. In Germany, service station income fell 30 per cent (Hertsgaard, 1999) and the financial losses and adverse publicity led Shell to cancel the sinking of the oil rig.

In the first decade of the twenty-first century, 'green consumerism' seems to have established itself as an integral part of the consumer market. The selling of selected produced services on their green, bio and eco credentials is common to many consumer markets. Although it is difficult to define exactly what green consumerism is, and therefore to measure it, consumer surveys of behaviour patterns, based upon environmental attitudes, indicate that environmental concern will play an increasing part in consumer behaviour in the future. For instance, when a sample of 2,000 people in Britain were asked if they had 'Selected one product over another because of environmentally friendly packaging/ formulation/ advertising', 36 per cent said they had in 1996 compared to 19 per cent in 1988 (Martin, 1997). If this survey were repeated now, it could be expected that the percentage figure would be higher.

Market research results in America also support evidence of a trend towards green consumerism. A survey by Michael Peters (MPG) of Americans in the summer of 1989 found that 53 per cent of the public had decided not to buy a product during the previous year, owing to concerns over the effects of the product or the packaging upon the environment. A survey carried out by the same firm in Canada found that approximately 66 per cent of the Canadian public would be more likely to buy a product that is recyclable or biodegradable, and approximately 60 per cent said they would be willing to pay more for such products (Cairncross, 1991). The growth in demand for organic or bio-farm products also supports the notion that people want to consume in a healthier, less environmentally damaging and more ethical way. The relatively high prices charged for organic farm produce in comparison to other produce reflect the willingness of a sizeable part of the consumer market to pay premium prices for 'environmentally sound' produce.

The willingness of people to pay more for products and services that are advertised as being environmentally sensitive does, however, offer the opportunity for consumer exploitation by companies. Claims that a company operates in an 'environmentally friendly way' are often difficult to substantiate, and some people may remain sceptical that a company's claim of employing green practices, and acting in an environmentally conscious way, is little more than a public relations exercise. There is a subsequent necessity for the regulation of environmental claims made by companies through independent bodies.

In the tourism sector, besides the International Standards Organisation (ISO) logo that is available to companies who implement 'environmental management systems', other green labels emphasising green credibility have appeared over recent years. One such scheme is the 'Green Globe', which was originally launched by the World Travel and Tourism Council

(WTTC) in 1994 (Hawkins, 1997). The WTTC is a trade association which incorporates many of the leading multinational travel and tourism corporations. The Green Globe logo is applicable to many of the different elements of the tourism system, including tour operators, transport services, visitor attractions, hotels, local communities and destinations. Any travel and tourism company can apply to become a Green Globe member as long as they want to be involved in sustainable tourism development. The advantage to organisations that are granted Green Globe accreditation is that the use of the logo should act as a signifier of environmental responsibility to the tourist who is influenced in their purchasing behaviour by environmental concerns.

> **THINK POINT**
>
> Do concerns over the natural environment influence your decisions of what you consume or buy? If yes, give a couple of examples. If no, why not?

Consumer trends and green tourism

The influence of green consumerism upon the tourism market is not easy to discern. At the end of the 1990s, commenting on green consumerism, Swarbrooke and Horner (1999: 198) remarked: 'One thing is clear – as the debate has developed, the term "green tourist" has not achieved the acceptance that the phrase "green consumer" has in general.' A decade later there is little evidence to suggest that this position has changed. This does not mean that consumers do not have concerns about the activities of the tourism industry. The interests of consumers in the environmental aspects of tourism were reflected as early as 1991 with the publishing of the book, *The Good Tourist*, which advised tourists how to act in a more environmentally responsible way. As Wood and House (1991) comment on the back cover:

> Thankfully now, the mistakes of the past are being recognised and a new breed of tourist is emerging. Travelling independently or in small groups, living as locals caring for the country they are visiting and its peoples, the Good Tourists take a long-term look at travelling and the world we live in.

Empirical research also supports the view that consumers were aware of the environmental impacts of the tourism industry by the late 1990s. Martin (1997) cites a survey carried out by MORI in the summer of 1995 in Britain, in which when asked 'How much damage to the environment do you think travel and tourism causes?', 64 per cent of the respondents replied that tourism caused some degree of damage to the environment. However, in the same survey, when asked about the damage to the environment caused by tourism in comparison to other industries, tourism ranked 14th out of 17 possible choices, behind other services such as road transport and domestic waste disposal. If a similar survey of public opinion about tourism's influence upon the environment were undertaken today, it would be interesting to gauge the extent to which there has been a shift in public opinion, especially given the media coverage of the contribution of aircraft emissions to global warming.

In the same survey, the opinion of the public about the efforts of the tourism industry to mitigate any of the harmful environmental impacts caused by it, in comparison to the efforts of other industries to reduce the negative environmental impacts their operations may be causing, was not reassuring for the sector. When presented with the statement, 'I'd like you to tell me how much you feel each industry is doing to reduce any harmful effects its activities might have on the environment', tourism ranked 16th out of 17 industries behind 'nuclear fuel manufacturing and reprocessing' and 'oil exploration and production' (Martin, 1997). However, awareness by the tourism industry that it has to be seen to be taking action on reducing its negative effects on the environment was demonstrated by recent publicity for a low-cost airline. Under the iconic image of a tree is printed the title: 'Care about the environment? We do too!' The advert then proceeds to make three key points: (i) that although aviation accounts for a small percentage of greenhouse gas emissions, the airline is not complacent; (ii) during the course of last year they removed from service 22 older aircraft at a cost of approximately US$550 million in an attempt to reduce emissions; (iii) because they fly more passengers per plane than a traditional airline, their emissions per passenger kilometre are 27 per cent less.

Although the effects environmental concerns will have on the future of the tourism market are difficult to predict, an evident trend over the last 15 years is the growth of a range of holiday types that infer greater awareness of the environment. Many terms have been used, often interchangeably, to describe these new forms of tourism including 'alternative', 'green', 'nature', 'sustainable', 'responsible' and 'ecotourism'. A common theme of these labels is that they are indicative of a more caring approach to how tourism should be developed and consumed. As Shackley (1996: 12) comments on new approaches to tourism:

> Terms such as environmentally friendly tourism, sustainable tourism, ecotourism, responsible tourism, low impact tourism are just a few

> among many in common use.... These designate low impact tourism
> programmes which might result in some form of sustainable benefits
> to the destination area.

Certainly, the idea of alternative types of tourism to mass tourism seems
to have found favour with a significant segment of the tourist market.
Besides the growth in environmental consciousness since the late 1980s,
the development of alternative forms of tourism can also be associated
with consumer over-familiarity with mass tourism, and a subsequent
desire for new types of holidays. This latter point means that the concept
of 'alternative tourism' can be interpreted in at least two ways: alternative
tourism as a form of more environmentally aware tourism; or alternative
tourism as types of tourism that are different to mainstream tourism
without necessarily being any less environmentally damaging.

So what are the criteria of alternative tourism? Whilst there is no
universal agreement on a definition of what alternative tourism actually
is (Brown, 1998), the differences of alternative tourism to mass tourism
are highlighted by Cater (1993: 85) as follows: 'Activities are likely to be
small scale, locally owned with consequentially low impact, leakages and
a high proportion of profits retained locally. These contrast with large-
scale multinational concerns typified by high leakages which characterise
mass tourism.' Utilising this definition, it is possible to highlight the
characteristics of alternative tourism which differentiate it from
mainstream tourism, as shown in Box 9.1.

Using these criteria, alternative tourism surpasses purely a concern for
the physical environment which typifies green tourism, to include
economic, social and cultural considerations. If the physical,
environmental and cultural dimensions of the environment are considered
in an integrated fashion, and tourism is developed with the characteristics
displayed in Box 9.1, then alternative tourism can be viewed as
synonymous with the concept of sustainable tourism development.

A kind of tourism that is often associated with the characteristics of
alternative tourism outlined above is 'ecotourism'. According to Shackley
(1996), it is a term invented by conservationists in the 1970s, whilst
according to Fennell (1999), the term can be traced back as far as 1965
to the work of Hetzer, who used it to explain the interaction between
tourists and the environments that they come into contact with. However,
like the concept of alternative tourism, there is no consensus about the
meaning of ecotourism (Goodwin, 1996; Fennell and Dowling, 2003).

Box 9.1

Characteristics of alternative tourism

- Small scale of development with high rates of local ownership
- Minimised negative environmental and social impacts
- Maximised linkages to other sectors of the local economy, such as agriculture, reducing a reliance upon imports
- Retention of the majority of the economic expenditure from tourism by local people
- Localised power sharing and involvement of people in the decision-making process
- Pace of development directed and controlled by local people rather than external influences.

The difficulty of trying to define ecotourism is referred to by Cater (1994: 3) thus:

> In particular, the term ecotourism is surrounded by confusion. Is it a form of 'alternative tourism' (furthermore, what is 'alternative tourism?')? Is it responsible (defined in terms of environmental, socio-cultural, moral or practical terms)? Is it sustainable (however defined)?

Waldeback (1995) recognises different dimensions and interpretations of what ecotourism can be, as shown in Box 9.2.

Evidence of the confusion over the term can be seen in the continued attempts to define what it is. For example, McLaren (1998: 97) comments:

> Ecotravel involves activities in the great outdoors – nature tourism, adventure travel, birding, camping, skiing, whale watching, and archaeological digs that take place in marine, mountain, island and desert ecosystems. Much of the travel is now called ecotourism, although critics argue that the definition of 'ecotourism' is so broad that almost any travel would qualify, as long as something green was seen along the way.

Box 9.2

Dimensions of ecotourism

Activity – tourism which is based upon experiencing natural and cultural resources

Business – tour operators who provide ecotourism tours

Philosophy – a respect for land, nature, people and cultures

Strategy – a tool for conservation, economic development and cultural revival

Marketing device – for promoting tourism products with an environmental emphasis

Handle – convenient umbrella name for a number of tourism-related concepts such as 'responsible or ethical travel', 'low-impact tourism', 'educational travel', 'green tourism' and so on

Symbol – of the debate about the relationship between tourism and the environment

Principles and goals – defining the symbiotic and sustainable relationship between tourism and the environment.

Source: Adapted from Waldeback (1995)

The danger of ecotourism being used as a marketing term in a disingenuous fashion to promote unsustainable forms of tourism is highlighted by Goodwin (1996), who comments upon its opportunist usage by the tourism industry, where the tag 'eco' has become synonymous with responsible consumerism. Hall and Kinnaird (1994: 111) also make the point about the problematic nature of defining ecotourism, stating that there is no consistent definition:

> This term [ecotourism] is used in a generic sense to cover tourism development which is sympathetic to, complements and/or is employed as a vehicle for, conserving and sustaining natural and cultural environments and their resources and which may encompass the domain of such terms as 'sustainable', 'green', 'soft', and 'alternative' tourism.

Holden and Kealy (1996: 60) add an explicit economic component in their definition:

> Implicit in all the definitions is respect or friendliness for the physical and cultural environment, i.e. developing a form of tourism that is non-damaging and non-degrading; subject to adequate and appropriate management controls; and that offers financial contributions for the protection of indigenous cultures and environments.

Goodwin (1996: 288) links this economic dimension directly to conservation of resources in his definition of ecotourism as:

> Low impact nature tourism which contributes to the maintenance of species and habitats either directly through a contribution to conservation and/or indirectly by providing revenue to the local community sufficient for local people to value, and therefore protect, their wildlife heritage area as a source of income.

This definition emphasises the economic aspect of ecotourism in conservation, and the involvement, instead of the exclusion of local communities in the conservation process. A comprehensive list of guiding principles for the development of ecotourism is set out in Box 9.3 from Wight (1994: 40).

The direct relevance of ecotourism to sustainability, following the principles of Wight (1994), is highlighted by Shackley (1996: 13), who states:

Ecotourism projects should meet the following criteria. They must:

- Be sustainable (defined as meeting present needs without compromising the ability to meet future needs)
- Give the visitor a unique and outstanding experience
- Maintain the quality of the environment.

Waldeback (1995) also suggests a range of goals that should be accommodated within the process of ecotourism development:

- Sustainable use
- Resource conservation
- Cultural revival and decolonisation
- Economic development and diversification
- Life enhancement and personal growth

Box 9.3

Guiding principles for ecotourism

- It should not degrade the resource and should be developed in an environmentally sound manner
- It should provide long-term benefits to the resource, to the local community and industry (benefits may be conservation, scientific, social, cultural, or economic)
- It should provide first-hand, participatory and enlightening experiences
- It should involve education amongst all parties – local communities, government, non-governmental organisations, industry and tourists (before, during and after the trip)
- It should encourage all-party recognition of the intrinsic values of the resource
- It should involve acceptance of the resource on its own terms, and in recognition of its limits, which involves supply-oriented management
- It should promote understanding and involve partnerships between many players, which could include government, non-governmental organisations, industry, scientists and locals (both before and during operations)
- It should promote moral and ethical responsibilities and behaviour towards the natural and cultural environment by all players.

Source: After Wight (1994: 40)

- Maximum benefits and minimal costs/impacts
- Learning about the natural culture and environment.

The confusion over what ecotourism actually is inevitably means that there is a lack of consistency in how it is implemented. Perhaps the most accepted definition is that of the International Ecotourism Society (2007): 'Responsible travel to natural areas that conserves the environment and improves the well-being of local people.'

However, the degree to which ecotourism will have any meaningful application for conservation is the extent that it is fed into tourism policy.

As Hall (2003: 21) comments: 'Ecotourism policy does not occur in a vacuum. Ecotourism policies are the outcome of a policy-making process which reflects the interaction of actors' interests and values in the influence and determination of the tourism planning and policy process.' This is an important perspective, reflecting the views of Wheeller, who since the early 1990s has been questioning the assumption that ecotourism will necessarily be environmentally beneficial. More recently (2005: 263) he comments:

> One of my main criticisms was that while ecotourism (or, indeed, sustainable tourism in its many guises) may, as a planning 'control', be fine in theory, it is useless in practice – primarily because it does not, and cannot confront the unfortunate harsh realities of human behaviour.

He may be correct, only time will tell. Certainly, stakeholders' and individuals' environmental ethics and values, as discussed in Chapter 2, are likely to be determinatory in the meaningfulness of alternative and ecotourism from a conservation perspective.

Who are the 'ecotourists'?

Based upon the principles and goals of ecotourism described by Wight (1994), Waldeback (1995), Shackley (1996) and the International Ecotourism Society (2007), it is evident that ecotourism places a much heavier emphasis upon conservation, education and ethics than does mass tourism. The emergence of a form of tourism that is based upon the premise of an ethical relationship with the environment is reflective of changes that are occurring in the consumer market for tourism. Poon's (1993) concept of the 'new tourist' was referred to in Chapter 2. According to Poon, 'new tourists' display a 'see and enjoy but do not destroy' attitude and do not assert that the 'west is best'. Other labels have also been given to the tourists who constitute this emerging environmentally conscious market segment, including the 'ethical tourist', 'environmentally responsible tourist', 'good tourist' and 'ecotourist' (Swarbrooke and Horner, 1999).

So who are the ecotourists? Just as there is no definitive definition of ecotourism, similarly there are different opinions on the characteristics of ecotourists, and who ecotourists actually are. Given the difficulty in defining ecotourism, it is not surprising that it is difficult to categorise the 'ecotourist', which leads to confusion over the type of behaviour that ecotourists could be expected to display. As Cater (1994: 76) comments:

> There is an inherent risk in assuming that the ecotourist is automatically an environmentally sensitive breed. Although small, specialist, guided groups of ecotourists may attempt to conform to this identity, the net has now been cast sufficiently wide to include less responsible behaviour.

It would therefore be mistaken to think that an 'ecotourist' will necessarily desire to be environmentally educated and have a minimal impact upon the environment. According to Mackay (1994), the tourist industry divides ecotourists into three distinct categories, of the big 'E', little 'E' and soft adventure types. The most popular group is the little 'E', in which tourists' environmental concerns are characterised by a wish to know that the hotel, airline or tour operator they intend to use has acceptable environmental standards. Big 'E' travellers are willing to travel into new, 'undiscovered' areas, and to accept the standards of accommodation and services offered by local people or camp in the wilderness. The 'soft adventurer' also wishes to visit wilderness areas, but wishes to visit them in comfort, however, without a sense of feeling that the nature or culture of the area they are visiting is being 'exploited' through tourism.

A further categorisation of tourists into different types relating to their level of interest in the environment is shown in Figure 9.1, based upon the work of Cleverdon (1999).The level of demand for each typology is reflected in the width of the base of each segment; that is, demand decreases upwards from the base of the pyramid to its apex. The model suggests that the largest segment of the tourist market, the 'Loungers', have a low level of interest in the environment beyond its providing pleasant surroundings. The focus of the holiday of this typology is likely to be based upon relaxing and enjoyment. The second typology of tourist, the 'Users', are interested in the environment having the special features that are required for the type of holiday they wish to pursue. The types of environment required for this typology are therefore specialised and limited. For example, typical activities associated with this group would include wildlife watching and downhill skiing, each of which needs an environment that possesses specialised features. The next typology, the 'Eco-aware', show an increasing interest in the environment, not for how they can use it but for its own sake. They have an interest in environmental issues connected with tourism, and would look for

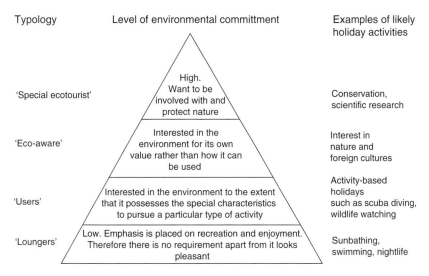

Figure 9.1 Types of tourist, based on their level of interest in the environment
Source: After Cleverdon (1999)

evidence of environmental commitment, and perhaps certification of a high level of environmental practice from the tourism companies and suppliers they use. Typical types of holiday activity will reflect an interest in knowing more about the nature and the culture of the destination they are visiting. The last group, the 'Specialist ecotourists', have a high level of commitment to the environment, to the extent that they want to actively protect it. This is reflected in the kinds of holiday they participate in, such as conservation holidays or scientific research.

Just as it is incorrect to talk about tourists as being one homogeneous group, Swarbrooke and Horner (1999) suggest that it is not possible to talk about the 'green tourist' as if they were one homogeneous group, preferring to refer to 'shades of green'. They suggest that the environmental commitment of tourists will be influenced by an amalgam of different factors, including: their awareness and knowledge of the issues associated with tourism and the environment; attitudes towards the environment in general; and the degree of fulfilment of other commitments in their life such as employment, housing and family needs. The different categories of green tourist and their associated environmental attitudes and actions are shown in Figure 9.2.

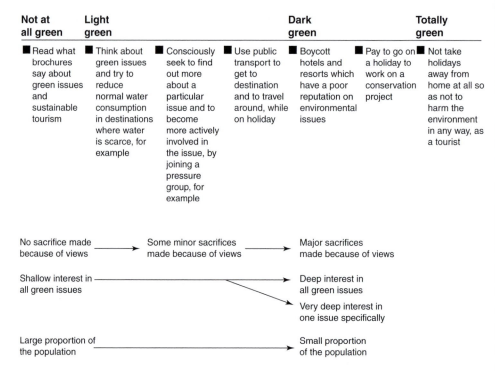

Figure 9.2 *Shades of green in tourism*

Source: J. Swarbrooke and S. Horner, *Consumer Behaviour in Tourism*, Oxford: Butterworth-Heinemann, p. 202 (1999)

Based upon empirical research in the USA, Wearing and Neil (1999) found that ecotourists have higher than average incomes and levels of education, and are also willing to spend more than normal tourists. In terms of their psychographic characteristics, Wearing and Neil report that they possess an environmental ethic, and are biocentric rather than anthropocentric in orientation.

In a study of Taiwanese ecotourists, Chang-Hung *et al.* (2004) found that there were five defined major themes of why people thought of themselves as being ecotourists. They were: (i) having a sense of environmental responsibility; (ii) having a strong interest in learning about nature; (iii) having a love of nature; (iv) participating in 'ecotourism activities', e.g. observing wildlife; and (v) visiting national parks and other natural areas.

However, the extent to which all ecotourists will have an ecocentric ethic has been challenged in research undertaken by Zografos and Allcroft (2007).

The diversity of environmental values and attitudes held by people thought of as ecotourists, because they display a propensity to visit areas of natural beauty, was unearthed by Zografos and Allcroft (ibid.). In a sample of visitors to an area of natural beauty in Scotland, four groups with varying mixes of 'environmental values' were identified. The largest segment (44.6 per cent) was termed the 'Disapprovers' because of their concern about the limits of natural resources and negative views on human attitudes and behaviour towards nature. The second largest segment (33.4 per cent), labelled the 'Scepticals', possessed a low level of concern for the limits of the Earth's resources and expressed scepticism in the ability of human skill and ingenuity to control nature. Conversely, the 'Approvers' (11.5 per cent) had high confidence in the ability of human skill to control nature and disagreed with criticisms of human attitudes towards nature. The smallest segment, the 'Concerners', refused to consider animal and human rights as equal but displayed a concern with the Earth and its limits.

> **THINK POINT**
>
> Referring to Figure 9.1 and 9.2, where would you locate yourself in terms of your 'level of interest in the environment' or 'shade of green'?

A rosy future? Ecotourism: product or principle?

The rationale for including alternative and ecotourism in the concluding chapter of the book is to reiterate the point made in the first chapter, that tourism does not take place in a void that is separate from the values, attitudes, processes and behaviour of society. It is observable that values, attitudes and behaviour do shift with time; for example, attitudes to child labour, racism, casual sex, smoking and drinking when driving have changed to varying degrees during the last century across different cultures. Similarly, the emergence of ecotourism and pro-poor tourism is perhaps reflective of a shift in values that are indicative of a wider concern about the environmental consequences of tourism at a global level.

However, similar to MacCannell's (1992) view of the creation of national and urban parks as an attempt to assuage guilt about the impact of human activity upon nature, Wheeller (1993a, 1993b, 2005) suggests that ecotourism is little more than a means for tourists to assuage guilt whilst in pursuit of their own short-term interests. Certainly, a major issue that faces the future of tourism's relationship with the natural environment is air travel. The choice to fly perhaps epitiomises the debate between individual self-interest and the greater good. It also illustrates the spatial discontinuity in the concept of sustainable tourism development, which continues to remain destination focused. The increased demand for air travel contradicts the notion of a sustainable form of tourism. Will the individual who holds a view that global warming is harmful and a consequence of human action be willing to forgo a flight and holiday and reduce their carbon footprint for the benefit of nature and future generations?

Alongside ethics, the future of tourism's relationship with the environment will also be determined by the wider political economy and existing hegemonic relationships. Fennell and Dowling (2003) observe that whilst different stakeholders (governments, private sector, NGOs, donor agencies, tourists and local communities) have an interest in ecotourism, money will drive the agenda, suggesting that some voices will be more powerful than others in deciding how tourism is developed.

Subsequently, how ecotourism will manifest itself is uncertain: will it lean towards a product or principle? This question was raised by Cater and Lowman (1994) in the early 1990s to illustrate the different interpretations of the term 'ecotourism'. Although this chapter has illustrated how ecotourism may be defined using a set of principles which reflect very closely the concept of sustainable development, yet there is also a distinct danger that ecotourism may be interpreted purely as another tourism 'product' to be sold in the marketplace. This may involve putting landscapes that have experienced little or no human interference directly in the marketplace for tourism. As Baroness Chalker, then Minister for Overseas Development for the United Kingdom government commented, one of the characteristics of ecotourism is 'attracting tourists to natural environments which are unique and accessible' (Cater and Lowman 1994: 90). Given that many of the most beautiful natural environments are located in rural areas of less developed countries, where often there are problems of underdevelopment and poverty, the economic incentives to be gained from ecotourism mean that there is a possible threat to the conservation of these areas.

This is a theme developed by Butler (1990) who questions, profoundly, the idea that forms of 'alternative' tourism offer an alternative approach to mass tourism. As he points out, the idea of 'alternative tourism' sounds good to everyone and is hard to disagree with, but it is so vague in its meaning that, like sustainable development, it can mean almost anything to anyone. Further, he continues that although mass tourism may be criticised a lot, many people still like to be mass tourists. It is therefore highly improbable that low-scale, alternative tourism will ever replace the mass tourism market. Butler is especially critical of the idea that one form of tourism can be promoted as the panacea to problems that have resulted from extensive and long-term tourism development, commenting: 'Mass tourism need not be uncontrolled, unplanned, short-term or unstable' (Butler, 1990: 40).

Harrison (1996) also questions the extent to which ecotourism and 'good tourists' are genuinely alternatives to mass tourism or whether they merely represent stages towards it. The danger of being complacent over ecotourism is also highlighted by De Alwis (1998: 232), who cites Robertson Collins the tourism veteran and conservationist as saying:

> Most of those who tag tourism 'ecotourism', after a while begin to believe that all they do is ecologically and people friendly. The customer who buys into an ecotourism labelled product also satisfies his or her conscience that they are not causing any damage to the nature or people that they visit.... Ecotourism today is big business. Products with an eco label are able to command higher prices in the marketplace. Similarly, there is also substantial funding for 'ecotourism' proponents by donors and well-meaning agencies, making it a lucrative consultancy based business as well. This situation has unfortunately led to the creation of the popular perception that 'tourism is bad' and 'ecotourism is good'.

De Alwis (1998) also makes the point that demand for ecotourism is market driven, as more people express an interest in nature, and the desire for independent travel has become stronger. The dangers of ecotourism are also elaborated by Shackley (1996), who suggests that the introduction of visitors into ecosystems that we barely understand can instigate long-term changes. Rogers and Aitchison (1998) draw attention to the point that, although the definitions of ecotourism sound attractive in theory, they do not address the more problematic issue of how to develop and promote ecotourism. They point out that ecotourism is often subject to the same market forces that led to the environmental problems associated with mass tourism, and that there are many different players involved in both the public and private sectors, leading to problems of co-ordination and planning of development. They add that ecotourism sites are diverse and at early stages of development, which limits the opportunities for comparative study, and that models of 'good practice' are lacking.

The ability to make financial profit from ecotourism is also emphasised by Mowforth and Munt (1998). In their view, not only does 'ecotourism' signal an interest in the 'environment' (*eco*logy) but it also indicates the ability to pay the high prices that such holidays command (*eco*nomic capital). Ecotourism is, as has been explained, a form of tourism that relies upon being alternative to the mainstream, and is therefore in one sense exclusive. The association of ecotourism with exclusivity is a theme emphasised by Wheeller in his extensively quoted work (1993a, 1993b), in which he renames ecotourism as 'egotourism'. Mowforth and Munt (1998) link the idea of 'egotourism' to being in a competition for distinction, uniqueness and differentiation, alternatively expressed as another form of 'conspicuous consumption'. Given the seeming willingness of ecotourists to pay premium rates to visit 'unspoilt' nature, there are strong economic and financial incentives for governments and business entrepreneurs to facilitate the use of the environment for ecotourism.

One of the great dangers in opening up new areas of the environment to ecotourism relates to the issue that unless development is strictly controlled, there is little reason to assume that the development cycle should be any different than for other forms of tourism. There has been and perhaps still is an inability to accept that ecotourists are a form of 'trail-blazer', opening up new areas to the hordes they disdain, unless there are strict development and planning controls (Burns and Holden, 1995). The application of planning controls limiting the number of ecotourists to a particular area may in reality prove difficult to implement and maintain, and may fluctuate in response to a country's economic situation and the policy of the government in power. This latter point is illustrated in Box 9.4 concerning the development of ecotourism in Belize.

Other criticisms have also been made of alternative forms of tourism like ecotourism, e.g. that they cannot provide a viable level of income and employment, and that education of the tourist into a form of behaviour that is respectful of the environment is a long-term process, for which the financing and logistics have not been thought through properly (Brown, 1998).

It is evident that alternative forms of tourism will not necessarily lead to tourism that is culturally and environmentally friendly. In a neo-liberal global economy, the regulation of market forces to allow a balanced use

Box 9.3

Ecotourism in Belize

One of the first countries to promote itself as an ecotourism destination was Belize in Central America, which is less than two hours' flying time from Miami in the USA. The country has a small land area of 23,000 square kilometres but contains a wide variety of vegetation types, including mangrove swamps, wetland savanna, mountain pine forests and tropical rainforests. Additionally, the coral reef off Belize's coast is the second longest in the world, after the Australian Barrier Reef. There are also several notable archaeological sites of the Mayan civilisation.

The attraction of ecotourism for the Belize government is to earn foreign exchange, and the government committed itself to balanced and environmentally sound tourism in the 1980s. The typical advertising slogans to promote Belize are 'Belize so natural', 'friendly and unspoilt', 'naturally yours'. The majority of tourists who come to Belize do not arrive on package tours, but their travel and accommodation are organised by foreign tour operators, and subsequently the economic leakage factor is likely to be high. The financial opportunities to be gained from ecotourism have led to rapid property inflation around the coastline and it is now estimated that 90 per cent of the coastline is under foreign ownership. US-based developers are building luxury resorts, with golf courses and marinas, and it is estimated that 65 per cent of the members of the Belize Tourism Industry Association are expatriates, mainly from the USA. However, the government has no plans to restrict foreign ownership of land as it requires the foreign investment.

The arrival of tourists at approximately 200,000 per annum is also beginning to have environmental consequences. Some of the coral in the Hol Chan Marine Reserve established in 1987 has become infected with black band disease, a form of algae which kills the coral, when it has already been broken. There has also been a decline in the numbers of conch and lobster owing to over-fishing, partly to satisfy tourism demand. Other examples of the impacts of tourism are more dramatic. In 1992, 'Eco-terrorism at Hatchet Cay' was the headline, in reference to a US resort owner who had attempted to blow up part of the coral reef to make his resort more accessible to visiting boats. Unfortunately, it would seem that ecotourism development in Belize has become characterised by the problems of mass tourism, including foreign exchange leakages, a high degree of foreign ownership of prime coastal land and tourism facilities, and environmental degradation. The government's desire for foreign exchange, combined with global market forces and trade liberalisation, means that the environment is becoming little more than a selling point for a form of mass ecotourism.

Sources: Cater (1992); Panos (1995); Mowforth and Munt (1998)

of environmental resources will be critical in determining the success of any alternative tourism policy. However, this may not be easy to achieve, especially for countries that are in need of foreign investment and foreign exchange. This means that for many developing countries, the global political economy, which incorporates international debt repayment and the domination of transnational corporations, places their environments under threat of exploitation for short-term need and profits. The use of environmental resources for ecotourism may, in terms of negative environmental consequences, prove to be little different to that of mass tourism, in fact perhaps even more detrimental as ecotourism environments are likely to be rich in biodiversity and highly susceptible to damage.

Concluding note

A central theme of this book has been that understanding the relationship that exists between tourism and the environment is concerned not just with the destinations that tourists go to but also the societies where they come from. It would seem that as societies become more economically developed, wealthier and urbanised, then the propensity to consume tourism increases. The projected levels of tourism demand of 1,600 million international tourist arrivals by 2020 (UNWTO, 2006C) suggest that tourism will become an even more significant feature of society and the global economy than it is at present.

This projected growth in tourism will provide both opportunities and threats for the natural environments and societies it interacts with. The economic opportunities offered by tourism mean it is increasingly likely to feature as a key part of many governments' economic policies, especially in developing countries. Importantly, by encouraging development and helping to eliminate poverty, tourism can assist in building a more sustainable future by meeting human needs, improving livelihoods and helping to conserve environments and wildlife, protecting them from threats such as poaching or potentially destructive forms of economic activity such as mining or logging. However, the achievement of this relies upon government policy reflecting a stronger commitment to an environmental ethic and a long-term perspective than has historically been the case.

The significant lesson of tourism development in the second half of the twentieth century was that although tourism can bring economic benefits, it can also contribute to the destruction of the natural environment. It can also be a cause of cultural change, for instance, changing the value systems of traditional societies by propagating consumerism and associated materialist values, and also displacing people from their traditional lands and denying them access to resources they require to meet their needs. Many of the negative effects of tourism have resulted from a *laissez-faire* approach to development, determined by free-market forces, in which the full environmental and social costs of tourism development have failed to be reflected. Enlightened government policy and planning for tourism in the future is essential, reflecting a balanced approach to how natural resources are used, based upon recognition that the natural environment holds a variety of other values alongside the purely monetary. For want of a better word, a more 'sustainable' approach to tourism development is required.

As a key stakeholder and catalyst to tourism development, the tourism industry will also need to continue to address its role and responsibilities to the environment. The tourism industry continues to display increasing signs of market maturity, with the acquisition and merging of companies into larger multinational conglomerates, which are increasingly being floated on stock exchanges as public companies. This trend is likely to continue as the market demand for tourism continues to increase. These multinational companies will subsequently have increasing influence in many destinations and markets around the world, and interact with the many locally owned SMEs, which continue to constitute the major part of the supply market in most tourism destinations. In an increasingly global business environment and marketplace, which facilitates the movement of international capital and the establishment of business operations in different countries, major international companies have an increasing responsibility to address ethical issues of their operations. Such issues will include their interaction with local communities, indigenous cultures and the natural environment. Practical initiatives could include actions such as strengthening supply linkages to locally owned businesses; helping to reduce the level of economic leakages; and increasing the economic benefits of tourism for members of the local community; and developing environmental management systems to cover all aspects of their operations.

However, there is no one model that can act as a blueprint of how tourism, the natural environment and local communities can interact in the most beneficial fashion in the future. Different destinations will have differing degrees of tolerance to tourism, based upon their own environmental characteristics, and the extent to which the natural and cultural environments have already been subjected to change by outside influences. Some tourism destinations, such as many of those on the western Mediterranean coastline, are already in a mature stage of development. For these areas, the major physical and cultural changes that tourism can bring have already been experienced, and in many of these areas efforts are being made to redress the negative effects of tourism. Other natural environments have yet to be 'discovered' by tourism; consequently, different destinations will have different tolerance levels to tourism. This is not to say that destinations in a mature stage of development are free from any further environmental damage from tourism, as continued expansion may push them past a further threshold level, where environmental management and technological measures cannot redress the balance. However, it is particularly in areas of the world where tourism is developing, or has yet to develop, that there is the potential for many environmental problems. As the pressure grows for kinds of tourism that are increasingly based upon the natural environment, including special ecosystems, wildlife and indigenous cultures, there is a requirement for strict planning controls to avoid unacceptable levels of environmental change.

A further issue in defining tourism's future relationship with the environment is that ultimately what is an acceptable or unacceptable level of environmental change and development will not be quantitatively fixed but determined by decision-makers in society. Tourism development and economic growth cannot take place without some degree of trade-off between the use of the environment and economic benefits. Ultimately the extent of this trade-off, and the amount of environmental change incurred and whom it benefits, will be a reflection of decision-makers' value systems and environmental philosophy. There is therefore a danger that where decisions are made by a small elite, there will be an inability to hear a wide range of views across the political and philosophical spectrums, and decisions will not be made on a consensual basis. There are subsequently likely to be clashes over the use of natural resources for tourism, particularly in situations where local people are denied access to resources that they have traditionally used to meet their needs. Tourism thus becomes viewed by many local people not as a constructive force for development but as a propagator of inequality. It is therefore necessary that the process of tourism development incorporates community participation, a principle agreed to by the majority of the world's governments when they became signatories of Agenda 21 in 1992.

Yet the future of tourism's relationship with the environment does not rely solely on what is happening in the tourism destination. Mass participation in international tourism is a feature of economically advanced countries, reflecting privilege, and is now an integral part of a post-modern consumer lifestyle. Whatever the motivations for participation in tourism, whether as a consequence of anomie and the need for ego-enhancement or the desire for escapism vis-à-vis the search for the authentic, tourism is a function of economically advanced and urbanised societies. The desire for tourism is a reflection of the quality of life experience of many people living in western society and those in societies that are becoming increasingly westernised. Empirical evidence would suggest that, whilst material wealth is higher than ever before in western countries, the quality of life in many of them has begun to decrease. For example, according to the Index of Social Health published by the Fordham Institute which measures well-being, America's social health declined from 73.8 out of a possible 100 in 1970 to 54 in 2004 (the latest year for which complete data is available) as homicides, suicides, the gap between the rich and the poor and drug use all increased (Fordham Institute for Innovation in Social Policy, 2007). Likewise, the pressures of the consuming life are also evident in Britain. In the newspaper article 'Britain in 2010: Rich but Far Too Stressed to Enjoy It', the point is made that the pressure of modern working patterns and the desire for wealth are fuelling family breakdowns, drug and alcohol dependency. Nearly a quarter of all men and women in Britain believed that they had sacrificed seeing their children grow up because of the pressures of work (Watson-Smyth, 1999).

What tourism represents to individuals will also be influential in determining the relationship it has with the surrounding environment. The extent to which it represents a search for escapism or authenticity, the experiences that are desired, and the amount of thought and planning devoted to it, will all be influential in determining how natural resources are used for tourism. As the world becomes increasingly mobile and more people seek to consume the experiences and rewards of tourism, a major question facing tourism planners is how increases in international and domestic tourism will be facilitated. Many of these extra tourists will be travelling by air transport with the subsequent effects upon the atmosphere from the emissions of carbon and nitrogen oxides. Yet it is particularly in tourism destinations where the pressures on the environment will be most intense, as this is where tourism is most concentrated.

Of particular concern is the increasing demand for nature-based tourism, which potentially could take tourists to some of the most ecologically fragile areas of the Earth. As already mentioned, alternative tourism such as ecotourism may in reality represent little more than the first stage of mass tourism. If tourists can go to space, which has already happened for a handful of the world's super-rich and will be likely to become more mainstream in the near future, no part of the Earth's environment can be considered to be inaccessible to tourism.

The relationship between tourism and the environment therefore faces many challenges in the future. Economic development and the social conditions of post-modern societies are a fertile breeding ground for tourism demand. The answer to a harmonious relationship in the future between tourism and the environment thus lies as much within the holistic philosophy of society and environmental attitudes as it does with a reductionist approach of finding technical solutions to environmental problems caused by tourism. The emergence of green consumerism since the late 1980s points to a realisation, at least by a part of the population, that present rates of consumption cannot continue without adverse environmental effects. Towards the end of the first decade of the twenty-first century, we are aware that many adverse environmental effects are the consequence of human actions, including global warming, ozone depletion, pollution of rivers and acid rain. There is increasingly a realisation, judging by the demand for more environmentally friendly products and ethical investment funds, that individual action can make a difference to environmental issues such as pollution and fair trade.

Some people are also increasingly disenchanted with the 'rat race' associated with 'comparative consumption', in which we work harder to buy more goods and services, to keep pace with our neighbours. The idea of 'downsizing' to a more simple lifestyle is an option that some people have chosen to take in the western world, perhaps indicative of wider changes to come. Conversely, a more pessimistic view would be that the majority of people will continue to act out of self-interest and in a myopic timeframe, unwilling to forgo lifestyle benefits and forms of enjoyment based upon a destructive use of natural resources, for the greater environmental good. The desire in the world for levels of material wealth enjoyed in western societies also dictates that global society will have a destructive relationship with the natural environment, unless a radically different development path can be found to the one that has been followed in the west. The increased consumption of goods and services means that tourism will bring changes to many natural environments around the world. The extent to which these can be determined as being positive or negative will be a reflection of the values held in society at that time and depend upon who is doing the judging.

Summary

- The emergence of green consumers in post-modern society represents a change in environmental values of a significant part of the market. The growth in demand for items such as organic foods, cosmetics that are free of animal testing, fair trade items and ethical investment funds suggests that ethical and environmental concerns are becoming more important for some people. The extent to which this consumer concern has materialised in the tourism market is limited. A key test of the depth of the environmental ethic will be the extent to which individuals are willing to forgo the benefits of air travel for the greater environmental good.
- There is a growing demand to visit environments that are regarded as being 'natural' and 'unspoilt'. Part of this demand is associated with the increasing levels of urbanisation and associated stress in society, which have removed people from contact with nature. One type of tourism that is increasingly talked about is 'ecotourism'. However, ecotourism is about much more than visiting nature, being based upon a set of principles advocating ethical approaches to tourism development and visitation. An alternative view of ecotourism is that it represents little more than a means for tourists to assuage guilt whilst in pursuit of their own short-term interests.
- The increasing demand for nature-based tourism is a source of concern. This concern is caused by the realisation that, without strict planning controls, there is little reason why what is commonly referred to as 'alternative tourism' should be anything more than the first step towards mass tourism. Many people also like to be mass tourists, and a key challenge for tourism planners in the future will be how to make mass tourism more sustainable.

Further reading

Now is the time to explore the bibliography!

Bibliography

Adam, D. (2007) 'Flights Reach Record Levels Despite Warnings over Climate Change', *Guardian*, May 9, p. 3.

Allaby, M. (ed.) (1994) *The Concise Oxford Dictionary of Ecology*, Oxford: Oxford University Press.

Ashley, C. and Mitchell, J. (2005) 'Can Tourism Accelerate Pro-Poor Growth in Africa?', London: ODI.

Ashley, C., Boyd, C. and Goodwin, H. (2000) *Pro-Poor Tourism: Putting Poverty at the Heart of the Tourism Agenda*, London: ODI.

Ashley, C., Roe, D. and Goodwin, H. (2001) *Pro-Poor Tourism Strategies: Making Tourism Work for the Poor: A Review of Experience*, London: ODI.

Association of Independent Tour Operators (2007) 'RT Guidelines', www.aito.co.uk, accessed March 23.

Attfield, R. (2003) *Environmental Ethics*, Cambridge: Polity.

Badger, A., Barnett, P., Corbyn, L. and Keefe, J. (1996) *Trading Places: Tourism as Trade*, London: Tourism Concern.

Baker, S., Kousis, M., Richardson, D. and Young, S. (eds) (1997) *The Politics of Sustainable Development: Theory, Policy and Practice within the European Union*, London: Routledge.

Barke, M. and France, L.A. (1996) 'The Costa del Sol', in Barke, M., Towner, J. and Newton, M.T. (eds) *Tourism in Spain: Critical Issues*, Wallingford: CAB International, pp. 265–308.

Barke, M. and Towner, J. (1996) 'Exploring the History of Leisure and Tourism in Spain', in Barke, M., Towner, J. and Newton, M.T. (eds) *Tourism in Spain: Critical Issues*, Wallingford: CAB International, pp. 3–34.

Barker, L.M. (1982) 'Traditional Landscape and Mass Tourism in the Alps', *Geographical Review*, 72 (4): 395–415.

Bart, J.M. van der Aa, Groote, P.D. and Huigen, P.P.P. (2005) 'World Heritage as NIMBY? The Case of the Dutch Part of the Wadden Sea', in Harrison, D. and Hitchcock, M. (eds) *The Politics of World Heritage: Negotiating Tourism and Conservation*, Clevedon: Channel View Publications, pp. 11–22.

Bartelmus, P. (1994) *Environment, Growth and Development: The Concepts and Strategies of Sustainability*, London: Routledge.

BAT (1993) *European Tourism Analysis*, Hamburg: BAT-Leisure Research Institute.

Bayswater, M. (1991) 'Prospects for Mediterranean Beach Resorts: an Italian Case Study', *Tourism Management*, 5: 75–89.

Becheri, E. (1991) 'Rimini and Co – the End of a Legend?: Dealing with the Algae Effect', *Tourism Management*, 12 (3): 229–35.

Beeton, S. (1997) 'Visitors to National Parks: Attitudes of Walkers to Commercial Horseback Tours'. Paper given at the Trebbi Conference, 6–9 July, Sydney, Australia.

Belle, N. and Bramwell, B. (2005) 'Climate Change and Small Island Tourism: Policy Maker and Industry Perspectives in Barbados', *Journal of Travel Research*, 44: 32-41.

Bird, B.D.M. (1989) *Langkawi: from Mahsuri to Mahathir: Tourism for Whom?*, Selangor: Malaysia Institute of Social Analysis.

Bocock, R. (1993) *Consumption*, London: Routledge.

Bodlender, J.A. and Ward, T.J. (1987) *An Examination of Tourism Investment Incentives*, London: Horwath & Horwath.

Body Shop, (1992) *The Green Book*, West Sussex: Body Shop International.

Boo, E. (1990) *Ecotourism: the Potentials and Pitfalls, Vol. 1*, Washington: World Wide Fund for Nature.

Boorstin, D.J. (1992) [1961] *The Image: A Guide to Pseudo-Events in America*, New York: Vintage Books.

Booth, D.E. (1998) *The Environmental Consequences of Growth*, London: Routledge.

Bowcott, O., Traynor, I., Webster, P. and Walker, D. (1999) 'Analysis: Green Politics', *Guardian*, March 11, p. 17.

Bowes, G. (2005) 'Tourism Hangs in the Balance', *Observer*, June 19, p. 4.

Bramwell, B. (2007) 'Opening Up New Spaces in the Sustainable Tourism Debate', *Tourism Recreation Research*, 32 (1): 1-9.

Brendon, P. (1991) *Thomas Cook: 150 Years of Popular Tourism*, London: Secker & Warburg.

Briguglio, L. and Briguglio, M. (1996) 'Sustainable Tourism in the Maltese Isles', in Briguglio, L., Butler, R., Harrison, D. and Filho, W.L. (eds) *Sustainable Tourism in Islands and Small States*, London: Pinter, pp. 161–79.

Brohman, J. (1996) 'New Directions in Tourism for Third World Development', *Annals of Tourism Research*, 23 (1): 48–70.

Brown, D.O. (1998) 'Debt-funded Environmental Swaps in Africa: Vehicles for Tourism Development?', *Journal of Sustainable Tourism*, 6 (1): 69–77.

Bruun, O. and Kalland, A. (1995) *Asian Perceptions of Nature: A Critical Approach*, Richmond: Curzon Press.

Budowski, G. (1976) 'Tourism and Conservation: Conflict, Coexistence or Symbiosis', *Environmental Conservation*, 3: 27–31.

Bull, A. (1991) *The Economics of Travel and Tourism*, London: Pitman.

Burac, M. (1996) 'Tourism and Environment in Guadeloupe and Martinique', in Briguglio, L., Butler, R., Harrison, D. and Filho, W.L. (eds) *Sustainable Tourism in Islands and Small States: Case Studies*, London: Pinter, pp. 63–74.

Burns, P. (1999) *An Introduction to Tourism and Anthropology*, London: Routledge.

Burns, P. and Holden, A. (1995) *Tourism: A New Perspective*, Hitchin: Prentice-Hall.

Butcher, J. (2003) *The Moralisation of Tourism: Sun, Sand ... and Saving the World?*, London: Routledge.

Butler, R. (1990) 'Alternative Tourism: Pious Hope or Trojan Horse?', *Journal of Travel Research*, 28 (3): 40–5.

Butler, R. (1993) 'Pre- and Post Impact Assessment of Tourism Development', in Pearce, D.W. and Butler, R.W. (eds) *Tourism Research: Critiques and Challenges*, London: Routledge, pp. 135–54.

Butler, R. (1997) 'The Concept of Carrying Capacity for Tourism Destinations: Dead or Merely Buried?', in Cooper, C. and Wanhill, S. (eds) *Tourism Development: Environmental and Community Issues*, pp. 11–22.

Butler, R. (1998) 'Sustainable Tourism – Looking Backwards in Order to Progress', in Hall, M.C. and Lew, A.A. (eds) *Sustainable Tourism: A Geographical Perspective*, Harlow: Longman, pp. 25–34.

Cairncross, F. (1991) *Costing the Earth*, London: The Economist Books.

Carley, M. and Spapens, P. (1998) *Sharing The World: Sustainable Living and Global Equity in the 21st Century*, London: Earthscan Publications.

Cater, E. (1992) 'Profits from Paradise', *Geographical Magazine*, 64 (3): 17–20.

Cater, E. (1993) 'Ecotourism in the Third World: Problems for Sustainable Tourism Development', *Tourism Management*, April: 85–90.

Cater, E. (1994) 'Ecotourism in the Third World – Problems and Prospects for Sustainability', in Cater, E. and Lowman, G. (eds) *Ecotourism: A Sustainable Option*, Chichester: Wiley, pp. 69–85.

Cater, E. and Lowman, G. (1994) *Ecotourism: a Sustainable Option*, Chichester: Wiley.

Chang-Hung, T.; Eagles, P.F.J. and Smith, S.L.J. (2004) 'Profiling Taiwanese Ecotourists Using a Self-definition Approach', *Journal of Sustainable Tourism*, pp. 149-68.

Christoplos, I. (2006) *Links between Relief, Rehabilitation and Development in the Tsunami Response: a Synthesis of Initial Findings*, London: Tsunami Evaluation Coalition.

Cioccio, L. and Ewen, M.J. (2007) 'Hazard or Disaster; Tourism Management for the Inevitable in Northeast Victoria', *Tourism Management*, 28 (1): 1-11

Clarke, J. and Critcher, C. (1985) *The Devil Makes Work: Leisure in Capitalist Britain*, Basingstoke: Macmillan.

Cleverdon, R. (1999) *Lecture Notes, Centre for Leisure and Tourism Studies*, London: University of North London.

Club Freestyle (1999) *Summer '99: Have It Your Way*, 2nd edn, London: Thomson Holidays.

Coccossis, H. and Parpairis, A. (1996) 'Tourism and Carrying Capacity in Coastal Areas: Mykonos, Greece', in Priestley, G.K., Edwards, J.A. and

Coccossis, H. (eds) *Sustainable Tourism: European Experiences*, Wallingford: CAB International, pp. 153–75.

Cohen, E. (1972) 'Towards a Sociology of International Tourism', *Social Research*, 39 (1): 164–89.

Cohen, E. (1979) 'A Phenomenology of Tourist Experiences', *Sociology*, 13: 179–201.

Cohen, E. (1995) 'Contemporary Tourism – Trends and Challenges: Sustainable Authenticity or Contrived Post-modernity?', in Butler, R. and Pearce, D. (eds) *Change in Tourism: People, Places, Processes*, London: Routledge, pp. 12–29.

Collin, P.H. (1995) *Dictionary of Ecology and Environment*, 3rd edn, Teddington: Peter Collin Publishing.

Collins (2004) *Geography Dictionary*, Glasgow: Harpercollins.

Coker, A. and Richards, C. (eds) (1992) *Valuing the Environment*, Chichester: Wiley.

Cooper, C., Fletcher, J., Gilbert, D., Wanhill, S. and Shepherd, R. (1998) *Tourism: Principles and Practices*, 2nd edn, Harlow: Longman.

D'Auvergne, B.D.E. (1910) *Switzerland in Sunshine and Snow*, London: T. Wener Laurie.

Dalen, E. (1989) 'Research into Values and Consumer Trends in Norway', *Tourism Management*, 10 (3): 183–6.

Dann, G. (1977) 'Anomie, Ego-Enhancement and Tourism', *Annals of Tourism Research*, 4 (4): 184–94.

Davidson, J. and Spearritt, P. (2000) *Holiday Business: Tourism in Australia since 1870*, Victoria: Melbourne University Press.

Davidson, R. (1993) *Tourism*, 2nd edn, London: Pitman.

De Alwis, R. (1998) 'Globalisation of Ecotourism', in East, P., Luger, K. and Inmann, K. (eds) *Sustainability in Mountain Tourism: Perspectives for the Himalayan Countries*, Delhi: Book Faith India, pp. 231–6.

De Costa, A. (2001) 'The Science of Climate Change', *The Ecologist*, Dec./Jan.: 10-16.

Department of the Environment (1991) *Tourism and the Environment: Maintaining the Balance*, London: HMSO.

Department for International Development (1997) *Tourism, Conservation and Sustainable Development: Comparitive Report*, Vol. 1, London: DFID.

Douthwaite, R. (1992) *The Growth Illusion*, Bideford, Devon: Green Books.

Doyle, T. and McEachern, D. (1998) *Environment and Politics*, London: Routledge.

Drumm, A. (1995) 'Converting from Nature Tourism to Ecotourism in the Ecuadorian Amazon'. Paper given at the World Conference on Sustainable Tourism, Lanzarote, April.

Dubois, G. and Ceron, J.P. (2006) 'Tourism/Leisure Greenhouse Gas Emission Forecasts for 2050: Factors for Change in France', *Journal of Sustainable Tourism*, 14 (2): 171-91.

Dumazedier, J. (1967) *Towards a Society of Leisure*, New York: Free Press.

Dwivedi, O.P. (2003) 'Classical India' in Jamieson, D. (ed.) *A Companion to Environmental Philosophy*, Oxford: Blackwell, pp. 21-26.

Eadington, W.R. and Smith, V.L. (1992) 'The Emergence of Alternative Forms of Tourism', in Smith, V. L. and Eadington, W.R. (eds) *Tourism Alternatives: Potentials and Problems in the Development of Tourism*, Philadelphia: University of Pennsylvania Press.

Eagles, P.F.J., McCool, S.F. and Haynes, C.D. (2002) *Sustainable Tourism in Protected Areas: Guidelines for Planning and Management*, Cambridge: IUCN.

Edington, J. and Edington, A. (1986) *Ecology, Recreation and Tourism*, Cambridge: Cambridge University Press.

Elliott, J.A. (1994) *An Introduction to Sustainable Development: The Developing World*, London: Routledge.

English Tourist Board (1991) *Tourism and the Environment: Maintaining the Balance*, London: English Tourist Board.

European Tourism Analysis (1993) *Ten Main Characteristics for Quality Tourism*, Hamburg: BAT-Leisure Research Institute.

Eurostat (1997) *Indicators of Sustainable Development*, Luxembourg: Eurostat.

Evans, G. (1993) 'Tourists Rush for Kill a Seal Pup Holiday', *Evening Standard*, July 5, p. 10.

Ezard, J. (1998) 'Ship Ahoy', *Guardian*, March 23, p. 8.

Farrell, B. (1992) 'Tourism as an Element in Sustainable Development: Hana, Maui' in Smith, V. L. and Eadington, W.R. (eds) *Tourism Alternatives: Potentials and Problems in the Development of Tourism*, Philadelphia: University of Pennsylvania Press, pp. 115–32.

Farrell, B.H. and Twining-Ward, L. (2003) 'Reconceptualising Tourism', *Annals of Tourism Research*, 31 (2): 274–95.

Faulkner, B. (2001) 'Towards a Framework for Tourism Destination Management', *Tourism Management*, 22 (3): 135–47.

Fennell, D.A. (1999) *Ecotourism: an Introduction*, London: Routledge.

Fennell, D. (2006) *Tourism Ethics*, London: Routledge.

Fennell, D. and Dowling, R. (eds) (2003) *Ecotourism Policy and Planning*, Wallingford: CAB International.

Font, X., Cochrane, J. and Tapper, R. (2005) *Pay per Nature View: Understanding Tourism Revenues for Effective Management Plans*, Netherlands: World Wide Fund for Nature.

Fordham Institute for Innovation in Social Policy (2007) *Index of Social Health*, New York: Fordham Graduate Centre.

Foster, J.B. (1994) *The Vulnerable Planet: A Short Economic History of the Environment*, New York: Cornerstone Books.

Franklin, A. (2003) *Tourism: An Introduction*, London: Sage Publications.

Friends of the Earth (1997) 'Atmosphere and Transport Campaign', www.foe.co.uk.

Gallup (1989) *American Express Global Travel Survey*, Princeton: Gallup Organization Inc.

Gamero, E. (1992) 'Legislation for Sustainable Tourism: Balearic Islands', in Eber, S. (ed.) *Beyond the Green Horizon*, Godalming: World Wide Fund for Nature.

Garman, J. (2006) 'If I were ... Aviation Minister', *The Ecologist*, pp. 23-24.

Gill, R. (1967) *Evaluation of Modern Economics*, New Jersey: Prentice Hall.

Gningue, A.M. (1993) 'Integrated Rural Tourism Lower Casamance', in Eber, S. (ed.) *Beyond the Green Horizon: a Discussion Paper on the Principles for Sustainable Tourism*, Godalming: World Wide Fund for Nature.

Goodall, B. (1994) 'Environmental Auditing: Current Best Practice' in Seaton, A.V., Jenkins, C.L., Wood, R.C., Deike, P.U.C., Bennett, M.M., Maclellan, L.R. and Smith, R. (eds) *Tourism: The State of the Art*, Chichester: Wiley, pp. 655–64.

Goodall, B. and Stabler, M.J. (1997) 'Principles Influencing the Determination of Environmental Standards for Sustainable Tourism' in Stabler, M.J. (ed.) *Tourism and Sustainability: Principles to Practice*, Wallingford: CAB International, pp. 279–304.

Goodpaster, K.E. in Werhane, P.H. and Freeman, R.E. (1998) *Encyclopedic Dictionary of Business Ethics*, Oxford: Blackwell, pp. 51–7.

Goodwin, H. (1996) 'In Pursuit of Ecotourism', *Biodiversity and Conservation*, 5: 277–91.

Goodwin, H. (2005) *Natural Disasters in Tourism*, Occasional Paper 1, International Centre for Tourism, University of Greenwich.

Gosling, D. (1990) 'Religion and the Environment', in Angell, D.J.R., Comer, J.D. and Wilkinson, M.L.N. (eds) *Sustaining Earth: Response to the Environmental Threat*, London: Macmillan, pp. 97–107.

Gossling, S. and Hall, M. (eds) (2006) *Tourism and Global Environmental Change: Ecological, Social, Economic and Political Interrelationships*, London: Routledge.

Goudie, A. and Viles, H. (1997) *The Earth Transformed: an Introduction to Human Impacts on the Environment*, Oxford: Blackwell.

Grylls, C. (2006) 'Return of the Wild', *The Geographical*, December, Royal Geographical Society, London.

Gunn, C. (1994) *Tourism Planning: Basic, Concepts, Issues*, 2nd edn, London/ Washington: Taylor & Francis.

Hall, C. and Lew, A. (eds) (1998) *Sustainable Tourism: a Geographical Perspective*, Harlow: Addison Wesley Longman.

Hall, D. (2006) 'Tourism and the Transformation of European Space', *ATLAS Reflections*, Arnhem, pp 11-26.

Hall, D. and Kinnaird, V. (1994) 'Ecotourism in Eastern Europe', in Cater, E. and Lowman, G. (eds) *Ecotourism: a Sustainable Option*, Chichester: Wiley, pp. 111–36.

Hall, M. (2003) from Fennell and Dowling

Hardin, G. (1968) 'The Tragedy of the Commons', *Science*, 162: 1243–8.

Harrison, D. (1996) 'Sustainability and Tourism: Reflections from a Muddy Pool' in Briguglio, L., Archer, B., Jafari, J. and Wall, G. (eds) *Sustainable Tourism in Islands and Small States: Issues and Policies*, London: Pinter, pp. 69–89.

Harrison, D. (1998) 'Whales under Stress as Man Crowds the Sea', *Observer*, October 18, p. 7.

Hawkins, R. (1997) 'Green Labels for the Travel and Tourism Industry – A Beginner's Guide', *Insights*, July, pp. A11–A15, London: English Tourist Board.

Heilbroner and Thurow, (1998).

Hertsgaard, M. (1999) *Earth Odyssey*, London: Abacus.

Hobsbawn, E. (1962) *The Age of Revolution*, London: Abacus.

Hogan, L., Metzger, D. and Peterson, B. (eds) (1998) *Intimate Nature*, New York: Ballantine.

Holden, A. (1991) 'Asian Dynasty', *Leisure Management*, 11 (12): 29–30.

Holden, A. (1998) 'The Use of Skier Understanding in Sustainable Management in the Cairngorms', *Tourism Management*, 19 (2): 145–52.

Holden, A. and Kealy, H. (1996) 'A Profile of UK Outbound Environmentally Friendly Tour Operators?', *Tourism Management*, 17 (1): 60–4.

Holdgate, M. (1990) 'Changes in Perception', in Angell, D.J.R., Comer, J.D. and Wilkinson, M.L.N. (eds) *Sustaining Earth*, Basingstoke: Macmillan, pp. 76–96.

Holloway, C. (1998) *The Business of Tourism*, 5th edn, Harlow: Addison Wesley Longman.

Hooker, C.A. (1992) 'Responsibility, Ethics and Nature', in Cooper, D.E. and Palmer, J.A. (eds) *The Environment in Question: Ethics and Global Issues*, London: Routledge, pp. 147–64.

House, J. (1997) 'Redefining Sustainability: a Structural Approach to Sustainable Tourism', in Stabler, M. (ed.) *Tourism and Sustainability: Principles to Practice*, Wallingford: CAB International, pp. 89–104.

Hudman, E. (1991) 'Tourism's Role and Response to Environmental Issues and Potential Future Effects', *Revue de Tourisme* [The Tourist Review]), 4: 17–21.

Hunter, C. (1996) 'Sustainable Tourism as an Adaptive Paradigm', *Annals of Tourism Research*, 24 (4): 850–67.

Hunter, C. and Green, H. (1995) *Tourism and the Environment: A Sustainable Relationship?*, London: Routledge.

IFTO (1994) *Planning for Sustainable Tourism: The Ecomost Project*, Lewes: International Federation of Tour Operators.

International Ecotourism Society (2007) 'About Ecotourism'.

IPCC (Intergovernmental Panel on Climate Change) (2007a) *Climate Change 2007: The Physical Science (Basic summary for Policymakers)*, Geneva: IPCC.

IPCC (2007b) *Climate Change 2007: Impacts, Adaptation and Vulnerability*, Geneva: IPCC.

Iso-Ahola, E.S. (1980) *The Social Psychology of Leisure and Recreation*, Iowa: Wm. C. Brown.

Ittleson, W.H., Franck, K.A. and O'Hanlon, T.J. (1976) 'The Nature of Environmental Experience' in Wagner, S., Cohen, B.S. and Kaplan, B. (eds) *Experiencing the Environment*, New York: Plenum Press, pp. 187–206.

Jamrozy, U. and Uysal, M. (1994) in Uysal, M. (ed.) *Global Tourist Behaviour*, New York: The Haworth Press, pp. 135–60.

Jenner, P. and Smith, C. (1992) *The Tourism Industry and the Environment*, London: The Economist Intelligence Unit.

Jowit, J. and Soldal, H. (2004) 'It's the New Sport for Tourists: Killing Baby Seals', *Observer*, October 3, p.3.

Keefe, J. (1995) 'Water Fights', *Tourism in Focus*, 17: 8–9.

Kinnaird, V., Kothari, U. and Hall, D. (1994) in Kinnaird, V. and Hall, D. (1994) (eds) *Tourism: A Gender Analysis*, John Wiley and Sons, Chichester, pp. 1-31.

Kirkby, S.J. (1996) 'Recreation and the Quality of Spanish Coastal Waters', in Barke, M., Towner, J. and Newton, M.T. (eds) *Tourism in Spain: Critical Issues*, Wallingford: CAB International, pp. 190–211.

Klemm, M. (1992) 'Sustainable Tourism Development: Languedoc and Roussillon', *Tourism Management*, June: 169–80.

Krippendorf, J. (1987) *The Holiday Makers*, Oxford: Heinemann.

Lai, K.L. (2003) 'Classical China' in Jamieson, D. (ed.) *A Companion to Environmental Philosophy*, Oxford: Blackwell, pp. 21-26.

Lanjouw, A. (1999) 'Mountain Gorilla Tourism in Central Africa', owner-mtn-forum@igc.apc.org.

Law, C. (1993) *Urban Tourism: Attracting Visitors to Large Cities*, London: Mansell.

Laws, E. (1991) *Tourism Marketing*, Cheltenham: Stanley Thornes.

Lea, J.P. (1993) 'Tourism Development Ethics in the Third World', *Annals of Tourism Research*, 20 (4): 701–15.

Lechte, J. (1994) *Fifty Key Contemporary Thinkers: From Structuralism to Postmodernity*, London: Routledge.

Leech, K. (2002) in Fox, C. (ed.) *Ethical Tourism: Who Benefits?*, London: Hodder & Stoughton, pp. 75-94.

Lencek, L. and Bosker, G. (1998) *The Beach: the History of Paradise on Earth*, London: Secker & Warburg.

Leopold, A. (1949) *A Sand Country Almanac*, Oxford: Oxford University Press.

Lickorish, L.J. and Jenkins, C.L. (1997) *An Introduction to Tourism*, Oxford: Butterworth-Heinemann.

Lister, R. (2004) *Poverty,* London: Routledge.

Liu, Z. (2003) 'Sustainable Tourism Development: A Critique', *Journal of Sustainable Development*, 11 (6): 459–75.

Lovelock, J. (1979) *Gaia: a New Look at Life on Earth*, Oxford: Oxford University Press.

Lyons, O. (1980) 'An Iroquois perspective', in Vecsey, C. and Venables, R. (eds) *American Indian Environments*, Syracuse: Syracuse Unitversity Press, pp. 171–74.

MacCannell, D. (1976) *The Tourist: a New Theory of the Leisure Class*, New York: Schocken Books.

MacCannell, D. (1989) *The Tourist*, 2nd edn, London: Macmillan.

MacCannell, D. (1992) *Empty Meeting Grounds: the Tourist Papers*, London: Routledge.

McCarthy, M. (2007) 'Green Groups Dismayed as Flights Soar to Record High', *Observer*, London, May 9, p. 11.

McCool, S.F. (1996) 'Limits of Acceptable Change: a Framework for Managing National Protected Area: Experiences from the United States'. Paper presented at the Workshop in Impact Management in Marine Parks, Kuala Lumpur, Malaysia, August 13–14.

Mackay, A. (1994) 'Eco Tourists Take Over', *The Times*, February 17.

McKie, R. (2006) 'Global Warming Threatens Scotland's Last Wilderness', *Observer*, p. 17.

McLaren, D. (1998) *Rethinking Tourism and Ecotravel*, Connecticut: Kumarian Press.

McMichael, P. (2004) *Development and Social Change: A Global Perspective*, 3rd edn, London: Sage Publications.

Marcel-Thekaekara, M. (1999) 'Poor Relations', *Guardian*, Saturday Review Section, February 27, p. 3.

Martin, A. (1997) 'Tourism, the Environment and Consumers'. Paper given at 'The Environment Matters' conference, Glasgow, April 30.

Maslow, A.H. (1954) *Motivation and Personality*, New York: Harper.

Mason, P. and Mowforth, M. (1996) 'Codes of Conduct in Tourism', *Progress in Tourism and Hospitality Research*, 2 (2): 151–68.

Mathieson, A. and Wall, G. (1982) *Tourism: Economic, Physical and Social Impacts*, Harlow: Longman.

Middleton, V. (1988) *Marketing in Travel and Tourism*, Oxford: Heinemann.

Mieczkowski, Z. (1995) *Environmental Issues of Tourism and Recreation*, Lanham, MD: University Press of America.

Mill, R.C. and Morrison, A.M. (1992) *The Tourism System: an Introductory Text*, 2nd edn, New Jersey: Prentice Hall.

Milne, S. (1988) 'Pacific Tourism: Environmental Impacts and their Management'. Paper presented to the Pacific Environmental Conference, London, October 3–5.

Mishan, E.J. (1969) *The Costs of Economic Growth*, Harmondsworth: Penguin.

Momsem, J.H. (1994) 'Tourism, Gender and Development in the Caribbean' in Kinnaird, V., Kothari, U. and Hall, D. (1994) (eds) *Tourism: A Gender Analysis*, Chichester: John Wiley & Sons, pp. 106–20.

Monbiot, G. (1995) 'No Man's Land', *Tourism in Focus*, 15: 10–11, London: Tourism Concern.

Mowforth, M. and Munt, I. (1998) *Tourism and Sustainability: New Tourism in the Third World*, London: Routledge.

Moynahan, B. (1985) *The Tourist Trap*, London: Pan Books.

Murphy, P. (1985) *Tourism: a Community Approach*, London: Routledge.

Murphy. P. (1994) 'Tourism and Sustainable Development', in Theobald, W. (1994) *Global Tourism: the Next Decade*, Oxford: Butterworth-Heinemann, pp. 274–90.

Naess, A. (1973) 'The Shallow and the Deep, Long-range Ecology Movement: a Summary', *Inquiry*, 16: 95-100

Nash, D. (1979) 'The Rise and Fall of an Aristocratic Tourist Culture', *Annals of Tourism Research*, Jan./March, pp. 63–75.

Nash, R.F. (1989) *The Rights of Nature: a History of Environmental Ethics*, Wisconsin: The University of Wisconsin Press.

Nicholson-Lord, D. (1993) 'Mass Tourism is Blamed for Paradise Lost in Goa', *Independent*, January 27, pp. 10–11.

O'Reilly (1986) 'Tourism Carrying Capacity: Concepts and Issues', *Tourism Management*, 8 (2): 254–8.

O'Riordan, T. (1981) *Environmentalism*, 2nd edn, London: Pion.

Osborn, D. and Bigg, T. (1998) *Earth Summit II: Outcomes and Analysis*, London: Earthscan Publications.

Page, J. (1999) Travel and Tourism, *Guardian* (weekend section), November 6, p. 102.

Page, S. (1995) *Urban Tourism*, London: Routledge.

Panos (1995) 'Ecotourism: Paradise Gained, or Paradise Lost', *Panos Media Briefing*, 14: 1–15.

Parviainen, J., Pysti, E. and Kehitys, S. (1995) *Towards Sustainable Tourism in Finland*, Helsinki: Finnish Tourist Board.

Pattullo, P. (1996) *Last Resorts: the Cost of Tourism in the Caribbean*, London: Cassell.

Pearce, D. (1993) *Economic Values and the Natural World*, London: Earthscan Publications.

Pearce, D., Markandya, A. and Barbier, E. B. (1989) *Blueprint for a Green Economy*, London: Earthscan Publications.

Pearce, P. (1988) *The Ulysses Factor: Evaluating Visitors in Tourist Settings*, New York: Springer-Verlag.

Pearce, P. (1993) 'Fundamentals of Tourist Motivation', in Pearce, D.G. and Butler, R.W. (eds), *Tourism Research: Critiques and Challenges*, London: Routledge, pp. 113–34.

Pepper, D. (1993) *Eco-socialism: From Deep Ecology to Social Justice*, London: Routledge.

Pepper, D. (1996) *Modern Environmentalism: an Introduction*, London: Routledge.

Pi-Sunyer, O. (1996) 'Tourism in Catalonia' in Barke, M., Towner, J. and Newton, M.T. (eds) *Tourism in Spain: Critical Issues*, Wallingford: CAB International, pp. 231–64.

Plog, S. (1974) 'Why Destination Areas Rise and Fall', *Cornell Hotel and Restaurant Quarterly*, Nov., pp. 13–16.

Ponting, C. (1991) *A Green History of the World*, London: Sinclair-Stevenson.

Poon, A. (1993) *Tourism, Technology and Competitive Strategies*, Wallingford: CAB International.

Porritt, J. (1984) *Seeing Green: the Politics of Ecology Explained*, Oxford: Basil Blackwell.

Reid, D. (1995) *Sustainable Development: an Introductory Guide*, London: Earthscan Publications.

Reuters (1999) 'Cape to Seek Sex Tourists', *Guardian*, September 20, p. 12.

Richardson, D. (1997) 'The Politics of Sustainable Development', in Baker, S., Kousis, M., Richardson, D. and Young, S. (eds) *The Politics of Sustainable Development: Theory, Policy and Practice within the European Union*, London: Routledge, pp. 43–60.

Rogers, P. and Aitchison, A. (1998) *Towards Sustainable Tourism in the Everest Region of Nepal*, Kathmandu: International Union for Conservation (IUCN).

Roe, D., Leader-Williams, N. and Dalal-Clayton, B. (1997) *Take Only Photographs: Leave Only Footprints*, London: International Institute for Environment and Development.

Rossetto, A.; Li, S.; and Sofield T. (2007) 'Harnessing Tourism as a Means of Poverty Alleviation: Using the Right Language or Achieving Outcomes?', *Tourism Recreation Research*, 32 (1): 49-58.

Roussopoulos, D.I. (1993) *Political Ecology*, Montreal: Black Rose Books.

Ryan, C. (1991) *Recreational Tourism: a Social Science Perspective*, London: Routledge.

Saarinen, J. (2006) 'Traditions of Sustainability in Tourism Studies', *Annals of Tourism Research*, 33 (4): 1121–40.

Sachs, J. (2005) *The End of Poverty: How We Can Make It Happen in Our Lifetime*, London: Penguin.

Salem, N. (1995) 'Water Rights', *Tourism in Focus*, 17: 4–5.

Saville, N.M. (2001) *Practical Strategies for Pro-Poor Tourism: Case Study of Pro-Poor Tourism and SNV in Humla District, West Nepal*, PPT Working Paper No. 3, London: DFID.

Scheyvens, R. (2002) *Tourism and Development: Empowering Communities*, Harlow: Pearson Education.

Scott, J. (2001) 'Gender and Sustainability in Mediterranean Island Tourism' in Ioannides, Y.; Apostolopoulos, E. and Sonninez, E. (eds) *Mediterranean Islands and Sustainable Tourism Development- Practices, Management and Policy, Continuum*, pp 87–107.

Scottish Office (1996) *National Planning Policy Guidelines for Skiing*, Edinburgh: Scottish Office.

Sen, A. (1992) *Inequality Reexamined*, Oxford, Russel Sage Foundations and Clarendon Press.

Sen, A. (1999) *Development as Freedom*, Oxford: Oxford University Press.

Shackley, M. (1995) 'The Future of Gorilla Tourism in Rwanda', *Journal of Sustainable Tourism*, 3 (2): 61–72.

Shackley, M. (1996) *Wildlife Tourism*, London: International Thomson Business Press.

Sharpley, R. (1994) *Tourism, Tourists and Society*, Huntingdon: Elm Publications.

Shaw, S. (1993) *Transport: Strategy and Policy*, Oxford: Blackwell.

Short, J.R. (1991*) Imagined Country: Society, Culture and Environment*, London: Routledge.

Simmons, I.G. (1993) *Interpreting Nature: Cultural Constructions of the Environment*, London: Routledge.

Simmons, M. and Harris, R. (1995) 'The Great Barrier Reef Marine Park', in Harris, R. and Leiper, N. (eds) *Sustainable Tourism: an Australian Perspective*, Oxford: Butterworth-Heinemann.

Simons, P. (1988) 'Apr ski le deluge', *New Scientist*, 1: 46–9.

Sinclair, T.M. (1991) 'The Tourism Industry and Foreign Exchange Leakages in a Developing Country: the Distribution of Earnings from Safari and Beach Tourism in Kenya', in Sinclair, T.M. and Stabler, M.J. (eds) *The Tourism Industry: an International Analysis*, Wallingford: CAB International, pp. 185–204.

Sinclair, T.M. and Stabler, M. (1997) *The Economics of Tourism*, London: Routledge.

Singer, P. (1993) *Practical Ethics*, 2nd edn, Cambridge: Cambridge University Press.

Slattery, M. (1991) *Key Ideas in Sociology*, Walton-on-Thames, Surrey: Thomas Nelson & Sons.

Smith, A.D. (2007) 'Melting Glaciers Will Destroy Alpine Resorts within 45 Years', *Observer*, London, January 14, p. 5.

Smith, C. and Jenner, P. (1992) 'The Leakage of Foreign Exchange Earnings from Tourism', in *Economist Intelligence Unit, Travel and Tourism Analyst* (3), London: Economist Intelligence Unit.

Smout, C. (1990) *The Highlands and the Roots of the Green Consciousness*, Perth: Scottish National Heritage.

Soane, J.V.N. (1993) *Fashionable Resort Regions: Their Evolution and Transformation*, Wallingford: CAB International.

Somerville, C., Rickmers, R.W. and Richardson, E.C. (1907) *Ski Running*, 2nd edn, London: Horace Cox.

Stabler, M. (forthcoming) 'Local Community Influences on the Management and Conservation of Natural Environments for Leisure and Tourism'. Discussion Paper in *Urban and Regional Economics*, Department of Economics, University of Reading.

Stabler, M. and Goodall, B. (1996) 'Environmental Auditing in Planning for Sustainable Island Tourism', in Briguglio, L., Archer, B., Jafari, J. and Wall, G. (eds) *Sustainable Tourism in Islands and Small States: Issues and Policies*, London: Pinter, pp. 170–96.

Stabler, M. and Sinclair, M.T. (1997) *The Economics of Tourism*, London: Routledge.

Stern Report (2006) *Stern Review on the Economics of Climate Change*, London: HM Treasury Office.

Stevens, T. (1987) 'Wigan Pier', *Leisure Management*, 7 (6): 31–4.

Stone, C.D. (1993) *The Gnat Is Older than Man: Global Environment and Human Agenda*, Princeton: Princeton University Press.

Swarbrooke, J. and Horner, S. (1999) *Consumer Behaviour in Tourism*, Oxford: Butterworth-Heinemann.

Todd, S.E. and Williams, P.W. (1996) 'Environmental Management System Framework for Ski Areas', *Journal of Sustainable Tourism*, 4 (3): 147–73.

Torres, R. and Momsen, J.H. (2004) 'Challenges and Potential for Linking Tourism and Agriculture to Achieve Pro-poor Tourism Objectives', *Progress in Development Studies*, 4 (4): 294-318.

Tourism Industry Association of Canada (1995) *Code of Ethics and Guidelines for Sustainable Tourism*, Ottawa: Tourism Industry Association of Canada.

Tourism Planning and Research Associates (1995) *The European Tourist*, 8th edn, London: Tourism Planning and Research Associates.

Tour Operators Initiative (2007) *Tour Operators Initiative for Sustainable Tourism Development*, www.toi.org, accessed March 29.

Towner, J. (1996) *An Historical Geography of Recreation and Tourism in the Western World: 1540–1940*, Chichester: Wiley.

Townsend, P. (1979) *Poverty in the United Kingdom*, Harmondsworth: Penguin.

TTG (1999) 'No Waste of Space', *Travel Trade Gazette*, p. 19, Special Edition (Day 3), World Travel Market, London.

Tunstall, S.M. and Penning-Rowsell, E.C. (1998) 'The English Beach: Experiences and Values', *The Geographical Journal*, 163 (3): 319–32.

Turner, K.R., Pearce, D. and Bateman, I. (1994) *Environmental Economics: an Elementary Introduction*, Hemel Hempstead: Harvester Wheatsheaf.

Turner, L. and Ash, J. (1975) *The Golden Hordes: International Tourism and the Pleasure Periphery*, London: Constable.

UN (United Nations) (2006) *The Millennium Development Goals Report: 2006*, New York: UN.

UNDP (United Nations Development Program) (1997) *Human Development Report*, New York: UNDP.

UNDP (2006) *Human Development Report*, New York: UNDP.

UNEP (United Nations Environment Programme) (1995) *Environmental Codes of Conduct for Tourism*. Technical Report, No. 29, Paris: UNEP.

UNEP (2000) *Natural Disasters*, unep.org/geo2000

UNEP (2004) *Economic Impacts of Tourism*, http://www.unepie.org/pc/tourism.

UNEP (2005) *Integrating Sustainability into Business: A Management Guide for Responsible Tour Operations*, Paris: UNEP.

UNESCO (2006a) *Partnerships for Conservation*, www.unesco.org.

UNESCO (2006b) *Sustainable Tourism*, www.unesco.org.

UNWTO (United Nations World Tourism Organization) (2004) Seminar on Sustainable Tourism Development and Poverty Alleviation, Madrid.

UNWTO (2006a) *Tourism Highlights 2006*, Madrid: UNWTO.

UNWTO (2006b) *Tourism and Least Developed Countries: A Sustainable Opportunity to Reduce Poverty*, Madrid: UNWTO.

UNWTO (2006c) *Annual Report*, Madrid: UNWTO.

UNWTO (2007a) *Sustainable Tourism for the Elimination of Poverty*, Madrid: UNWTO.

UNWTO (2007b) *Increase Tourism to Fight Poverty – New Year Message from UNWTO*, Madrid: UNWTO.

Urry, J. (1990) *The Tourist Gaze: Leisure and Travel in Contemporary Societies*, London: Sage.

Urry, J. (1995) *Consuming Places*, London: Routledge.

Vardy, P. and Grosch, P. (1999) *The Puzzle of Ethics*, London: Fount.

Veblen, T. (1994) [1899] *The Theory of the Leisure Class*, 3rd edn, London: Penguin.

Vidal, J. (1994) 'Money for Old Hope', *Guardian*, January 7, pp. 14–15.

Vidal, J. (2007a) 'Vast Forests with Trees Each Worth £4,000 Sold for a Few Bags of Sugar', *Guardian*, April 11, p. 3.

Vidal, J. (2007b) 'China Could Overtake US as Biggest Emissions Culprit by November', *Guardian*, April 27, p. 13.

Visser, N. and Njuguna, S. (1992) 'Environmental Impacts of Tourism on the Kenya Coast', *Industry and Environment*, 15 (3): 42–51.

Wainwright, M. (2007) 'Respect for Wordsworth 200 Years on with Daffodil Rap', *Guardian*, April 11, p. 6.

Waldeback, K. (1995) *Beneficial Environmental Sustainable Tourism*, Vanuatu: BEST.

Ward, L. (1998) 'Quality of Life Gets a Higher Profile', *Guardian*, November 24, p. 2.

Wathern, P. (ed.) (1988) *Environmental Impact Assessment: Theory and Practice*, London: Routledge.

Watson-Smyth, K. (1999) 'Britain in 2010: Rich but Far Too Stressed to Enjoy It', *Independent*, September 15, p. 8.

WCED (World Commission on Environment and Development) (1987) *Our Common Future*, Oxford: Oxford University Press.

Wearing, S. and Neil, J. (1999) *Ecotourism: Impacts, Potentials and Possibilities*, Oxford: Butterworth-Heinemann.

Weston, J. (ed.) (1997) *Planning and Environmental Impact Assessment*, Harlow: Addison Wesley Longman.

Wheeller, B. (1993a) 'Sustaining the Ego?', *Journal of Sustainable Tourism*, 1 (2): 23–9.

Wheeller, B. (1993b) 'Willing Victims of the Ego Trap', *Tourism in Focus*, 9: 10–11.

Wheeller, B. (2005) 'Ecotourism/Egotourism and Development' in Hall, C.M. and Boyd, S. (eds) *Nature-Based Tourism in Peripheral Areas*, Clevedon: Channel View Publications, pp. 263–72.

White, L.W. (1967) 'Historic Roots of our Ecological Crisis', *Science* 155: 1203–07.

Whitt, L.A., Roberts, M., Norman, R. and Grieves, V. (2003) 'Indigenous Perspectives' in Jamieson, D. (ed.) *A Companion to Environmental Philisophy*, Oxford: Blackwell Publishing, pp. 3-20.

Wight, P. (1994) 'Environmentally Responsible Marketing of Tourism', in Cater, E. and Lowman, G. (eds) *Ecotourism: A Sustainable Option*, Chichester: Wiley, pp. 39–53.

Wight, P. (1998) 'Tools for Sustainability Analysis in Planning and Managing Tourism and Recreation in a Destination', in Hall, C. and Lew, A. (eds) *Sustainable Tourism: a Geographical Perspective*, Harlow: Addison Wesley Longman, pp. 75–91.

Williams, P.W. and Gill, A. (1994) 'Tourism Carrying Capacity Management Issues', in Theobald, W. (ed.) *Global Tourism: the Next Decade*, Oxford: Butterworth-Heinemann, pp. 174–87.

Williams, S. (1998) *Tourism Geography*, London: Routledge.

Willis, I. (1997) *Economics and the Environment: a Signalling and Incentives Approach*, St. Leonards: Allen & Unwin.

Wood, K. and House, S. (1991) *The Good Tourist*, London: Mandarin Paperbacks.

Wood, L. (1998) 'Quality of Life Gets a Higher Profile', *Guardian*, November 24, p. 2.

World Bank (2005) *Tsunami: Impact and Recovery: Joint Needs Assessment*, Washington: World Bank.

World Bank (2007) 'Poverty Drops Below 1 Billion, says World Bank', www.worldbank.org/data/wdi

World Guide (1997/8) *The World Guide: A View from the South*, Oxford: New Internationalist Publications.

World Tourism Organization (1991) *International Conference on Travel and Tourism Statistics*, Madrid: WTO.

World Tourism Organization (1992) *Tourism Carrying Capacity: Report on the Senior-Level Expert Group Meeting held in Paris, June 1990*, Madrid: WTO.

World Tourism Organization (1998a) *Tourism: 2020 Vision* (Executive Summary Updated), Madrid: WTO.

World Tourism Organization (1999a) *Tourism Highlights: 1999*, Madrid: WTO.

World Tourism Organization (1999b) 'Global Code of Ethics for Tourism', www.world-tourism.org/pressrel/CODEOFE.htm.

World Tourism Organization (2003) *Climate Change and Tourism*, Madrid, WTO UNWTO.

World Travel and Tourism Council (2007) *Progress and Priorities 2007*, London: WTTC. Worthington, S. (1999) 'Green Slogans Are Whitewash', *Evening Standard*, October 7, p. 23.

Zimmermann, F.M. (1995) 'The Alpine Region: Regional Restructuring Opportunities and Constraints in a Fragile Environment', in Montanari, A. and Williams, A.M. (eds) *European Tourism: Regions, Spaces and Restructuring*, Chichester: Wiley.

Zografos, C. and Allcroft, D. (2007) 'The Enviromental Values of Potential Ecotourists: A Segmentation Study', *Journal of Sustainable Tourism*, 15 (1): 44–65.

Index

Abu Simbel temples 180–1
acid rain 71, 73, 93, 95, 250
Adivasi people 91, 133
'adventurers' 54, 238
advertising of tourism 36–42
aesthetic pollution 96
Agenda 21 150–1, 168, 248
agricultural revolution 30
agro-chemicals 66
air pollution 75, 93–5, 101
air travel 15–16, 84, 93–4, 116, 219–20,
    241–2, 251
airport construction 81–4, 95
Aitchison, A. 243–4
Allcroft, D. 240–1
'allocentrics' 52–3, 56
Alpine region 30, 34–5, 74–5, 95, 216
alternative tourism 232–3, 242–4, 250–1
Amboseli National Park 121
American Airlines 7
American Express 16
Andorra 67
animism 60–1
Annan, Kofi 139–40
anomie 44, 46, 249
anthropocentrism 27, 59, 61–2, 120
Antigua 86
arrivals of tourists 20–3, 129
Ash, J. 68
Ashley, C. 138, 144
Association of Independent Tour Operators 195
Attfield, R. 26
auditing, environmental 195–9, 202, 209
Australia 17, 35
Austria 35
'authentic' experiences and cultures 49–52, 249
Azores, the 40–1

Baden-Baden 13
Bahamas 89
Baiae 11

Baker, S. 156
Balearic Isles 165, 179, 181
Bali 19
Baltimore 3, 100
bank holidays 15
Barbados 215
Barke, M. 69
Bartelmus, P. 110
Bath 13
beaches and beach holidays 22–3, 32–4,
    78–80, 85–6, 93
Becheri, E. 92
behaviour patterns of tourists 55–8, 64, 90,
    101, 206–7, 244
Belize 244–5
Belle, N. 215, 221
Bhutan 190
Biarritz 67
Bigg, T. 150
The Body Shop 228
Bookchin, Murray 157
Boorstin, D.J. 2, 44–6
Boulding, K.E. 66
Bourdieu, Pierre 43
bovine spongiform encephalopathy
    (BSE) 69
Bramwell, B. 161, 215, 221
Brendon, P. 12
Brent Spar oil rig 228
Briguglio, L. and M. 95–6
British Airways 7
Brundtland Report (1987) 71, 128, 148–52,
    156, 161, 164
Bruun, O. 26
Buddhism 28–9
budget airlines 54
Bull, A. 4
Burac, M. 96
Burns, P. 90–1, 164
business tourism 3
Butler, R. 160, 162, 190, 242–3

Cairncross, F. 117, 120, 227
Canada 62, 81, 185, 204–5, 229
capitalism 59–60, 63–4, 155–6
car travel 15, 93–5
carbon emissions 70, 72, 93–4, 101, 213–14, 219, 224
'careers' of tourists 46–7
Caribbean region 17, 22, 68, 71, 74–5, 81, 89, 92, 136, 210, 215, 218
Carlton Travel 7
Carrick-a-Rede 36–7
carrying capacity 161, 187–91; definition of 187
Carson, Rachel 66, 71
Carthage 11
Cater, E. 233, 238, 242
Catlin, George 178
Centre for Future Studies 183
Chalker, Linda 242
Chang-Hung, T. 240
Chernobyl 68, 71
China 21, 29, 94, 213–14, 224
chloro-fluorocarbons 227
Christianity 26–7
Cleverdon, R. 238
climate change 8, 36, 71–2, 126, 210–20, 227; definition of 211–12; influence on tourism 215–18, 224; tourism's contribution to 219–20
Club Freestyle 57
Club of Rome 67, 71
coastal areas 31–4, 64, 96, 216
Coccossiss, M. 158–60
codes of conduct 63, 203–9; criticisms of 206–9
Cohen, E. 44–5, 48–52
collective consumption 114
Collins, Robertson 243
commodity prices 106–7
common pool resources 74–6
community involvement in development 161, 164–5, 183–4, 204, 223
competition for resources 83–8
Concord, Mass. 59
Congo, Democratic Republic of 173
conservation 28, 63, 70, 98–100, 124–6, 148, 152, 161, 208
Conservation International (CI) 125
conspicuous consumption 43, 64, 244
consumer trends 230–7
consumerism 42, 70, 226–30, 247, 250–1
contingent valuation 121, 123
Cook, Thomas 16, 36–9

coral reefs 76–80, 89, 186, 218
Creutzfeldt-Jakob Disease 69
cultural changes resulting from tourism 90–1
Cumae 11

Dalen, E. 53–4
Dann, G. 46
Daoism 28
Davidson, J. 3
Davos 34
De Alwis, R. 243
'debt for nature' swaps 124–5
Defoe, Daniel 31–2
Delphi 67
'demonstration effect' of tourism 90
Department for International Development (DFID) 70, 123, 129, 146
Department of the Environment 163
Descartes, René 60
developing countries 4–5, 42, 63, 70, 90–1, 102, 106–13, 124, 128–9, 134–5, 138, 145–6, 172, 179, 208, 246
DINKS 54
Donegal 189
Douthwaite, R. 108–9
Dowling, R. 242
Doyle, Sir Arthur Conan 35
Doyle, T. 154, 156
'dreamers' 54
'drifters' 51–2
Drumm, A. 89
Durkheim, Emile 44

Earth Summit (Rio, 1992) 72, 151
ecocentrism 153–6, 164
eco-fascism 63
eco-feminism 158
ecology, *deep* and *shallow* 59–60, 154–7
ECOMOST project 165
economic analysis of tourism 108–13
economic growth 105–7, 149, 152, 155–6, 248; costs of 108, 126; through tourism 107, 110
'economisers' 54
eco-socialism 157–8
ecosystems 75–83, 212
ecotourism 63, 70, 72, 113, 135, 232–46, 250–1; definition of 233–5
Ecotourism Society 69, 72, 203
ecotourists 237–41
Ecuador 89
El Niño 78
Elliott, J.A. 152

Elyne, M. 35
enclave tourism 112
Engels, Friedrich 32
environment, the: definition of 25–9, 63;
  human behaviour in relation to 88–91;
  promotion as a tourist attraction 36–42;
  services provided for society by 103–4;
  tourists' experience of 55–8
environmental auditing 176, 195–200, 203, 209
environmental concerns 66–9, 227, 230–2,
  251
environmental impact analysis 174–5, 192–4
environmental impacts 67–73; negative 73–9;
  positive 97–101
environmental management systems (EMSs)
  197–8
environmental problems 28, 58–9, 145, 248
environmentally-adjusted domestic product 110
escapism 44–6, 249
ethical concerns 5, 39, 58–65, 69, 219
European Union 168, 193
exotic destinations 38–9, 86
expenditure on tourism 21–2
experience-orientated tourists 48–51
'experimental' mode of tourism 49, 51
'explorer' tourists 51–3
externalities 113–17; definition of 114;
  negative 115, 209

Farrell, B. 150, 188
Faulkner, B. 223
Fennell, D. 232, 242
Finland 198–9, 202
Fordham Institute 249
Fordist production 6
foreign exchange 107–8
foreign investment 172–4
France 15, 36, 68, 80–1
France, L.A. 69
Franco, Francisco 4, 18–19, 107, 181
Franklin, A. 5
French Riviera 11–12, 18
Friends of the Earth 83

General Agreement on Tariffs and Trade
  (GATT) 172–3
General Agreement on Trade in Services
  (GATS) 172–3
genetically-modified crops 69–72
Ghana 125
Gill, R. 12
global warming 8, 36, 40, 68, 71–2, 75–8, 93–4,
  101, 117, 149, 210–11, 214–15, 227, 250

Goa 74, 85–6
Goa Foundation 69, 86
Goldsmith, E. 68
golf courses 85–8, 93
Good, R. 35
Goodall, B. 203, 208
Goodwin, H. 234–5
gorillas 99–100
Goudie, A. 76
government policies for tourism 19–20, 108,
  140–1, 171–6, 208, 246–7
La Grande Motte 81
Granada 89
Grand Tour 11–13, 18, 29–30
Great Barrier Reef 76, 183–6, 192
Grecotels 200–2
Greece 20, 67
Greek civilisation 12
green activism 69–72, 226–30, 250–1
green consumerism 70–1,
226–230
green economics 118
Green Globe logo 229–30
green tourism 230–40
greenhouse gases 213, 219, 224
'greening' of tourism organisations 202
Greenpeace 71, 96, 228
Grenier, P. 35
gross domestic product 108–10, 126
gross national product 108–10, 126
guide books 12
Gunn, C. 7

habitus 43–6
Hall, C. 177, 237
Hall, D. 234
Hardin, G. 76, 118–19
Harrison, D. 164, 243
Hawaii 68, 76
health tourism 12–13, 32–5
heritage attractions 101; see also World
  Heritage Sites
Hilton International Hotels 7
Himalayan region 206–7, 221
Hinduism 28
Hobsbawm, E. 14
Holden, A. 101, 164, 235
Holloway, C. 11–12
Horizon Travel 18
Horner, S. 230, 239
House, J. 160–1
Human Development Index 134
human rights 63, 69

Humla 141–2
Hunter, C. 158–60
hunting 39, 62

identity, sense of 43
images of tourism and tourist destinations 67, 223–4
India 29, 213
Indian Ocean tsunami (2004) 220–4
indigenous peoples and cultures 27–8, 44, 74, 87
'indulgers' 54
Industrial Revolution 14, 23, 30–1, 35, 64, 213
information technology 16
Intergovernmental Panel on Climate Change 71–2
International Air Transport Association (IATA) 84
International Energy Agency 213–14
International Federation of Tour Operators 165
International Monetary Fund 174
International Standards Organisation 229
internet resources 16, 54
intra-generational equity 156, 161
Islam 29
Iso-Ahola, E.S. 55
Italy 67–8
Ittleson, W.H. 55–6

Jamaica 81
Jamrozy, U. 46
Jenkins, C.L. 5–6
Jenner, P. 95, 112
Jowit, J. 62

Kalland, A. 26
Kant, Immanuel 60
Katrina (hurricane, 2005) 220–1
Kealy, H. 235
Kehoe, Mike 62
Kenya 74, 77–8, 87–90
Kenyatta, Jomo 124
Kinnaird, V. 136, 234
Kiribati 40
Korea, Republic of 193
Krippendorf, J. 7, 58
Kyoto Agreement (1997) 72

labelling schemes 229–30
Lake District 31
landscape, perceptions of 29–36, 42
landslides 81–3
land-use planning 184–94, 208–9

Langkawi 85, 87, 174
Lanzarote 49–50, 97
Law, C. 100
Laws, E. 8
leakage of expenditure 111–13
Leech, K. 63
Leopold, Aldo 59
Lew, A. 177
Lewis, R. 36
Lickorish, L.J. 5–6
limits of acceptable change (LAC) 190–1
Limits to Growth report (1972) 67, 71
Lister, R. 132–3
litter 89
Liu, Z. 161
Liverpool 3, 100
local cultures, contact with 49–51
Lowman, G. 242
Lunn, Sir Henry 16, 36

Maasai people 77, 87
MacCannell, D. 44–6, 177–8, 241
McCool, S.F. 190–1
McEachern, D. 154, 156
Mackay, A. 238
McLaren, D. 233–4
Madagascar 124
Madikwe Nature Reserve 136–7
Malaysia 20, 174–5, 181
Malta 74, 95
Manchester 32
Maori people 27–8
Marbella 19
Marcel-Thekaekarka, M. 91
Martin, A. 230–1
Martinique 96
Maslow, Abraham 43
mass tourism 6–7, 10, 16–21, 31, 36, 51, 68, 71, 232, 243, 249–51
Mathieson, A. 4, 88, 187–8
Mayan civilisation 27
Mediterranean region 17, 22–3, 32, 75, 94, 138, 210, 218, 224, 248
Melbourne 14
Midland Railway 37, 39
Mieczkowski, Z. 95, 98, 114–15
MILKIES 54
Mill, R.C. 8
Millennium Development Goals (MDGs) 129–31, 136, 139–40, 145
Milne, S. 67
Mishan, E.J. 67, 114
Momsen, J.H. 145

Montpelier 12
Moore, Henry 61
Morocco 19
Morrison, A.M. 8
motivations of tourists 46–8, 63–4
mountain regions 30, 34, 36, 64, 81–3, 96
Mowforth, M. 43, 160, 206–8, 244
Muir, John 177
multinational corporations 112, 247
multiplier effects 111, 182
Munt, I. 43, 160, 206–8, 244
Murphy, P. 6, 178–9

Naess, Aerne 59
Naples 11
Nash, D. 13
Nash, R.F. 27, 62
National Environmental Policy Act
  (US, 1969) 192
national goals 171–2
national parks 63, 121–4, 175–9, 185, 208,
  241; costs and benefits of 180
natural disasters 220–4; tourism's response to
  223–4
nature, perceptions of 26–36, 59–60
Neil, J. 240
'neotenous tourism' (Hunter) 160
Nepal 83, 87, 141–3
Netherlands Development Organisation (SNV)
  70, 129, 141
'new tourism' 10, 54, 237
Nice 13, 35, 81
Nicholson-Lord, D. 86
noise pollution 95–6, 116
nomadic societies 30
NorSafari 62
Norway 62
nuclear power and nuclear weapons 67–8, 78

O'Riordan, T. 154, 171–2
oil spills 66, 71
opportunity costs 84
organically-produced food 70, 72, 251
Organisation for Economic Co-operation and
  Development (OECD) 68, 71, 115
Orwell, George 100–1
Osborn, D. 150
over-fishing 62–3, 78
Oxfam 70
ozone depletion 68, 71, 75, 94, 114, 149,
  227, 250

package holidays 6, 16, 18, 56, 112

Page, J. 16
Page, Stephen 8
Parpairis, A. 158–60
Parviainen, J. 197
Pattaya 58
Pearce, David 109, 115, 120–1, 124
Pearce, P. 46–7
Penning-Rowsell, E.C. 33–4
Pepper, D. 60, 153
personality characteristics of tourists 52
Peters, Michael 229
phenomenology of tourist experiences 48–9
Philippines, the 124
philosophy of the environment 59–61
Pigou, Arthur 106, 117
pilgrimage 11
Plog, S. 52–3
political ecology 161
'polluter pays' principle 115–17
pollution 75, 91–7, 101, 149, 250; costs of
  105, 110, 113
Poon, A. 6–10, 23, 54, 237
population growth 152, 157
Porritt, J. 109
poverty 107, 128–46, 152, 169, 221; *absolute*
  and *relative* 130, 132; definition of 130–4,
  146; tourism's role in alleviation of 134–46
poverty lines 132–3, 146
pre-leakage factors 112–13
price of travel 16–17
private sector role in environmental protection
  194–202, 209
product-led tourism 159
pro-poor tourism (PPT) 138, 140, 144,
  146, 241
protected areas 176–80, 185, 208
psychocentrics 52–3
psychographics 52
psychological aspects of tourism 46–7
public goods 114, 117–18
Puteoli 11

quality of life 108–9, 156, 169
quality tourism 41–2, 121–2

railway systems 15, 33–9
Raitz, Vladimir 18
regeneration, urban 3, 100, 102
Reid, D. 153
religious belief systems 26–9
repatriation of profits 112
resource depletion 105, 109–10
Richardson, D. 151

Rio conference (1992) *see* Earth Summit
Rio Tinto Zinc (RTZ) 164–5
Rogers, P. 243–4
Romagna coast 92, 98
Roman Empire 11–13, 61
Romantic movement 30–1, 40, 64
Rome 11, 67
Roosevelt, Theodore 39
Roussopoulos, D.I. 157–8
Russia 21
Rwanda 99–100

Saarinen, J. 161, 170
Sachs, J. 130–2
safari parks 77, 88
St Gervais 35
St Lucia Wetlands 164–5
St Moritz 34–5
Salem, N. 84
Salou 92–3, 98
San Francisco 14
Scheyvens, R. 136
Scotland 30, 74, 89, 96–7, 165, 217, 241
seal culling 62–3
seaside resorts 17, 32–3
Sen, Amartya 133–4, 144
Senegal 165–6
sex tourism 91
Shackley, M. 89, 122, 231–2, 235, 243
Sharpley, R. 54–5
Shaw, S. 116–17
Shell (company) 228
Sheraton Hotels 7
Short, J.R. 30
Simmel, Georg 43
Simmons, I.G. 27, 61, 83
Sinclair, T.M. 108
Singapore 101
Singer, P. 27
skiing and ski facilities 34–6, 70, 81–2,
    96–7, 216
small and medium-sized enterprises (SMEs)
    221, 247
Smith, Adam 12
Smith, C. 95, 112
Smout, C. 30
social costs 114,116, 247
social differentiation and status 43–6
social ecology 157–8
social exclusion 133
Social Health, Index of 249
sociologial perspectives on tourism 43–6
Soldal, H. 62

spa towns 13
space tourism 23
Spain 4, 16–19, 32, 68–71, 107, 181
Spinoza, Benedict 60
spirituality 27–30
Stabler, M. 108, 203, 208
standards, environmental 116–17
standards of living 106–9, 156, 172
STEP (Sustainable Tourism for the Elimination
    of Poverty) initiative 139–40, 146
stewardship of the environment 27, 98, 164–5
Stone, J.V.N. 124, 172
structural adjustment programmes 174
structuralist model of tourism 160–1
sustainability, concept of 160–3
sustainable development 28–9, 70, 72, 98,
    148–59, 169, 242; different perspectives on
    153–8; meaning of 150–2; origins of
    148–51; radical approaches to 156–7;
    strong and *weak* forms of 155–6, 159
sustainable tourism 113–14, 139–40, 150,
    158–70, 237, 251; definition of 165;
    Globe'90 goals of 162–3; guiding
    principles for 163; indicators of 167–8
Swarbrooke, J. 230, 239
Switzerland 35
Sydney 100
System of Integrated Environmental and
    Economic Accounting 110

Tahiti 67
taxation, environmental 116–17, 219–20
technocentrism 153–4
Tenerife 57
Thailand 91, 107, 193
Thomas, William 12
Thoreau, Henry 59
Three Mile Island 67, 71
time-consciousness 14–15
Titanic 16–17
Torremolinos 18–19
Torres, R. 145
Torrey Canyon 66, 71
tour operators 16, 39–40, 44, 52, 70, 183
Tour Operators' Initiative (TOI) for sustainable
    tourism development 195–6, 209
tourism: benefits to destination areas from
    107–13, 182–4; as a form of consumption
    42–7; growth in demand for 10–21, 246;
    influence of climate and nature on demand
    for 21–3; and sustainability 158–62; as a
    system 7–10, 23, 25, 83, 226; types of 3
Tourism Concern 69, 72, 206

tourism industry, definition of 5–7
Tourism Planning and Research Associates 45
Touristik Union International (TUI) 7, 199
tourists, types of 48–55
Towner, J. 12, 32–3
Townsend, Peter 133
travel-cost method (TCM) of valuing
  environments 121
trickle-down effect 107
tsunami see Indian Ocean tsunami
Tunstall, S.M. 33–4
Turner, K.R. 120–1
Turner, L. 68
Twining-Ward, L. 150

ultra poverty 132
United Nations 129, 132, 149, 158, 189;
  Development Programme (UNDP) 125, 130,
  134, 141–3; Educational, Scientific and
  Cultural Organisation (UNESCO) 175,
  180–1, 196; Environment Programme
  (UNEP) 176, 194, 196, 220; Statistical
  Commission 3; World Tourism Organisation
  see World Tourism Organisation
'unspoilt' environments 41–2, 244, 251
urban tourism 100–2
urbanisation 14, 30–2, 43–5, 64, 251
Urry, J. 29–33
utility, concept of 104
Uysal, M. 46

Vail, Colorado 70
Veblen, Thorstein 43–4
Venice 101
Vichy 13
Viles, H. 76

Wadden Sea 184
Waldeback, K. 233–6
Wall, G. 4, 88, 187–8
Walton, J.K. 13

water pollution 92–3
water resources 84–6
Wearing, S. 240
welfare, economics of 106–8
Weston, J. 192
wetlands 80–1
whale-watching 57
Wheeller, B. 237, 241, 244
White, L. 26
Whitelegg, J. 84
Wigan 100–1
Wight, P. 188, 235
Wild, M. 36
wildlife, viewing of 77, 88–9
  'wildscape' 13, 34–8
Williams, A. 22–3
Williams, S. 184–5
willingness to pay 120–3, 126, 229
Willis, I. 114
winter sports 16, 34–5, 216
Wordsworth, William 31
World Bank 107, 132, 140, 146, 174
World Commission on Environment and
  Development see Brundtland Report
World Conservation Strategy (1980) 148, 151
World Heritage Sites (WHSs) 175, 180–4
World Tourism Organisation 3–4, 63, 110,
  129, 139, 146, 176, 185, 187, 196, 215
World Trade Organisation 70, 173
World Travel and Tourism Council 5,
  229–30
World Wide Fund for Nature (WWF) 70, 120,
  124, 165, 203

Yosemite 177–9
YUPPIES 54

Zambia 124
Zimmermann, F.M. 95
Zografos, C. 240–1
zoning 184–7